Association for Women in Mathematics Series

Volume 1

More information about this series at http://www.springer.com/series/13764

Association for Women in Mathematics Series

Focusing on the groundbreaking work of women in mathematics past, present, and future, Springer's Association for Women in Mathematics Series presents the latest research and proceedings of conferences worldwide organized by the Association for Women in Mathematics (AWM). All works are peer-reviewed to meet the highest standards of scientific literature, while presenting topics at the cutting edge of pure and applied mathematics. Since its inception in 1971, The Association for Women in Mathematics has been a non-profit organization designed to help encourage women and girls to study and pursue active careers in mathematics and the mathematical sciences and to promote equal opportunity and equal treatment of women and girls in the mathematical sciences. Currently, the organization represents more than 3000 members and 200 institutions constituting a broad spectrum of the mathematical community, in the United States and around the world.

Kathryn Leonard • Sibel Tari
Editors

Research in Shape Modeling

Los Angeles, July 2013

Springer

Editors
Kathryn Leonard
Department of Mathematics
California State University Channel Islands
Camarillo, CA, USA

Sibel Tari
Department of Computer Engineering
Middle East Technical University
Ankara, Turkey

ISSN 2364-5733 ISSN 2364-5741 (electronic)
ISBN 978-3-319-16347-5 ISBN 978-3-319-16348-2 (eBook)
DOI 10.1007/978-3-319-16348-2

Library of Congress Control Number: 2015937479

Mathematics Subject Classification: 68-T45, 68-U10, 65-D18, 97-M60, 68-T10

Springer Cham Heidelberg New York Dordrecht London

Printed on acid-free paper

Springer is part of Springer Science+Business Media (www.springer.com)

Preface

In July 2013, a group of around 30 women researchers in shape modeling attended the Women in Shape Modeling (WiSh) workshop at the Institute for Pure and Applied Mathematics (IPAM) in the University of California Los Angeles (UCLA) campus to begin research collaborations on some of the major problems in shape analysis and modeling. Participants from North America, Asia, Europe, and the Middle East lived, ate, slept, and worked together for one week at IPAM, then continued their collaborations from a distance. This proceedings volume contains preliminary results from those collaborations and related work. We look forward to future papers on these topics as participants continue their contributions to the growing body of work on mathematical shape modeling.

WiSh participants worked on one of the four research questions below.

1. *Team Leaders: Luminita Vese, Sibel Tari.* Simultaneous spectral and spatial analysis of shape, investigating a new distance-like shape operator from the spectral point of view, adopting signatures developed in the spectral literature, and solving similar symmetry detection problems. We also develop a connection to image segmentation and registration using the yet unclear connection of the new operator to the Ambrosio-Tortorelli functional.
2. *Team Leaders: Aasa Feragen, Megan Owen.* Dimensionality reduction and visualization of data in tree-spaces, studying dimensionality reduction in shape spaces where the shapes have the structure of a tree, such as classes of anatomical trees like airways and blood vessels, medial axes of 2D shapes, or phylogenetic trees. We develop techniques for low-distortion embedding into open books and hyperbolic spaces whose geometric structure is similar to that of tree-space.
3. *Team Leaders: Kathryn Leonard, Erin Wolf Chambers.* Geometric shape segmentation, exploring shape segmentation from a Gestalt perspective, using information from the Blum medial axis of edge fragments in an image. We combine existing edge saliency measures together with medial data to increase support for or against hypothesized edge interpolation and develop techniques for considering related appearance cues.

4. *Team Leaders: Marie-Paule Cani, Raphaelle Chaine.* Representing and editing self-similar details on 3D shapes, studying shape deformation and editing techniques, such as elongating or compressing parts of a shape while maintaining local style, copy-pasting details from one shape to another, or changing the scale of details without changing lower resolution geometry. All are essential for interactive shape design. Blending properties of implicit surfaces make them good candidates for solving this problem. We explore the extension of multi-resolution analysis to these surfaces and their deformations, enabling us to characterize repetitive self-similarities, and develop methods for filtering details out and generating them again after low resolution shape editing. An extension studies multi-resolution editing of animated shapes.

It has been an honor to work with this exceptional group of women and with the efficient staff at IPAM. We are grateful to IPAM, the National Science Foundation, Microsoft Research, and the National Geospatial Agency for providing funding for the workshop, and to the UCLA Department of Mathematics for sharing space with us. In addition, the Association for Women in Mathematics (AWM), and President-Elect Kristin Lauter in particular, played a key role in encouraging me to organize a research collaboration workshop in conjunction with AWM. These workshops are now a regular part of AWM's offerings, and AWM activities at the annual Joint Mathematics Meetings center on research topics emerging from collaboration workshops.

Special thanks is due to my coeditor, Sibel Tari, who provided useful insights during the workshop and invaluable help during the preparation of this manuscript, and to Stacey Beggs at IPAM who worked creatively to solve several unexpected issues during the course of the workshop.

California, USA Kathryn Leonard
May 2014

Contents

Introduction

Modeling shapes is one of the most thought-provoking challenges in computational imaging and perception. Let us consider, for example, a deceptively simple task of comparing two shapes. Which properties of the shape should be the basis for comparison if one desires to imitate *common sense* or *intuitive nearness*? This question is hard to answer because such nearness is not absolute; it depends on the context, e.g., other shapes in the same environment. A property that holds a group of shapes together in one context may be an irrelevant detail or noise in another context. Imagine a collection of circles and triangles drawn using a rich variety of creative brushstrokes. By filtering out the brush effect as a useless detail, a perception system can recover two categories, the circle and the triangle. The brush effect – useless detail in one scenario – might, however, be the key feature holding a wide variety of contour drawings together in another scenario containing a wide range of basic shape contours drawn using regular pen and a zigzag brush. In this case, the two brush types naturally characterize the two categories. Note, however, that the goal of comparing shapes is not merely to construct equivalence relations on a given set of shapes; continuous measures of context-dependent nearness and right shape models that support them are what one needs. Formulating salient features of shapes, selectively removing detail, constructing flexible coordinate frames and shape spaces equipped with appropriate metrics, and computing in-betweens are all parts of the attempt to capture the continuum of relations among shapes. Several chapters in the book directly address these topics.

Two contributions (Chapters 8 and 9) by the M.P. Cani and R. Chaine team are on saliency and detail. The material in both chapters heavily rests on medial representation, a timeless paradigm in shape modeling. Medial representation is the main paradigm in Chapters 6 and 7, too, respectively contributed by the K. Leonard and E. W. Chambers and the S. Tari and L. Vese teams. In both chapters, the problem of computing medial representations from complex images is tackled.

Computing shape representations from complex images is important because shapes in the form of point clouds, surfaces, or characteristics functions are not readily available; they must be extracted using discontinuities in images. This is not an easy task. In order to make the process of shape extraction immune to noise, texture, contrast variations, so on and so forth, images need to be conditioned, i.e., irrelevant details need to be suppressed and missing information due to weak contrast or an occlusion must be recovered.

Intriguingly, conditioning of images requires prior knowledge on the nature of shapes being investigated. This opens up a whole new set of challenges, the most notably a need for communication between the processes of shape understanding and image filtering. The earlier process of image filtering is a necessary step before shape extraction. Nevertheless, its success heavily depends on the proper choice of prior knowledge of the nature of shapes in the image. Defining mid-level priors in the form of generic regularities and/or high-level priors in the form of shape categories for filtering and shape extraction is a fundamental problem of which solution rests on defining shapes via functions of certain regularity that can be differentiated and incorporated into multi-objective optimization problems involving both region and boundary terms. Chapter 1, a contribution by the S. Tari and L. Vese team tackles the problem of automatically selecting high-level shape priors when shapes are represented implicitly by real valued functions (fields) defined on the shape domain. The applications of representing shapes implicitly via fields are further investigated in two more contributions (Chapters 2 and 3) by the members of the same team.

Applications drive theoretical development. Medical and biological applications have been a major source of motivation in shape research. In Chapter 4, continuum mechanics models are explored for the analysis of the cortical shape. In Chapter 5, the A. Feragen and M. Owen team models the space of trees that stem from anatomical shapes. Indeed, discrete computational structures such as trees and graphs are indispensible tools in shape research. They are commonly employed to express relationships among shape components.

Put together, the chapters in the book cover an entire spectrum in shape analysis starting from raw images ending with shape-related decisions.

Ankara, Turkey Sibel Tari

Chapter 1
Automatic Prior Shape Selection for Image Segmentation

Weihong Guo, Jing Qin, and Sibel Tari

Abstract Segmenting images with occluded and missing intensity information is still a difficult task. Intensity based segmentation approaches often lead to wrong results. High vision prior information such as prior shape has been proven to be effective in solving this problem. Most existing shape prior approaches assume known prior shape and segmentation results rely on the selection of prior shape. In this paper, we study how to do simultaneous automatic prior shape selection and segmentation in a variational scheme.

1.1 Introduction

Image segmentation has many important applications in object recognition, machine learning, medical imaging, etc. In medical imaging for instance, segmentation of anatomical structures is used to help in diagnosis, surgical planning and evaluation. Intensity based image segmentation methods can be classified into region based, edge-based and a combination of these two. Using image intensity information alone however may not lead to desired results when the image to be segmented has significant signal loss, poor image contrast and missing boundaries. Prior shape based approaches are more effective in these cases. Most existing shape based approaches assume the shape prior is given and a misleading prior shape might lead to wrong segmentation. We use sparse optimization to automatically select

W. Guo (✉)
Department of Mathematics, Case Western Reserve University, Cleveland, OH, 44106, USA
e-mail: wxg49@case.edu

J. Qin
Department of Mathematics, University of California, Los Angeles, CA, 90095, USA
e-mail: jxq@ucla.edu

S. Tari
Department of Computer Engineering, Middle East Technical University, Çankaya, Turkey
e-mail: stari@metu.edu.tr

K. Leonard, S. Tari (eds.), *Research in Shape Modeling*, Association for Women in Mathematics Series 1, DOI 10.1007/978-3-319-16348-2_1

prior shapes from a shape library and simultaneously segment images. The proposed variational approach is able to automatically and adaptively select prior shape which in turn guides segmentation. It is especially beneficial when there are objects with multiple shapes to segment.

The rest of the paper is organized as follows: Sect. 1.2 introduces the proposed model. Numerical results are presented in Sect. 1.3. Conclusion is drawn in Sect. 1.4.

1.2 Model Description

In this section, we start by reviewing the Ambrosio-Tortorelli approximation of the Mumford-Shah model, then we describe how to apply it to form the shape library. Lastly, we present the proposed segmentation model.

1.2.1 Ambrosio-Tortorelli Approximation of Mumford-Shah Segmentation Functional

Given an image $g(x)$ defined on an open and bounded set $\Omega \subset \mathbb{R}^2$ satisfying $g \in L^\infty(\Omega)$, Mumford and Shah [1] proposed the following functional for image segmentation

$$F^{MS}(u,S) = \int_{\Omega/S} \left(\alpha|\nabla u|^2 + \beta|u-g|^2\right)dx + \mathcal{H}1(S).$$

where $\mathcal{H}^1$ is the Hausdorff 1-dimensional measure in $\mathbb{R}^2$, i.e.,

$$\begin{aligned}\mathcal{H}^1(S) &= \sup_{\delta>0} H^1_\delta(S) = \lim_{\delta\to 0} H^1_\delta(S)\\ &= \liminf_{\delta\to 0}\left\{\sum_{i=1}^{\infty}(\mathrm{diam}(U_i))^d : \bigcup_{i=1}^{\infty} U_i \supseteq S, \mathrm{diam}(U_i) < \delta\right\}.\end{aligned}$$

The functional is optimized in a weak sense and can be approximated by [2]

$$G^{AT}_\rho(u,v) = \int_\Omega \left[\rho|\nabla v|^2 + \alpha v^2|\nabla u|^2 + \frac{(v-1)^2}{4\rho} + \beta|u-g|^2\right]dx.$$

Then the image segmentation is to find a piecewise C^1 function $u(x)$ and a function $v(x)$, such that $v(x) \to 1$ as $\rho \to 0$ in the $L^2(\Omega)$-topology, i.e.,

$$\lim_{\delta\to 0}\int_\Omega |v-1|^2 dx = 0.$$

Neither Mumford-Shah model nor its Ambrosio-Tortorelli approximation can work well for images with missing or occluded edge information. Shape prior is required in this case to obtain a complete segmentation. We use sparse optimization to search for prior shapes that adapt to images automatically.

1.2.2 Shape Descriptor Library

We start by reviewing edge strength functions presented in [3] to form a library. Then we will describe how to use these functions to form our shape libraries. These edge strength functions have distance function look and provide richer information than the binary silhouette images (see Fig. 1.1). For notational simplicity, we use the same notation to interchangeably represent a matrix and its vectorized version. For the rest of the paper, we consider discrete models. For instance,

$$\int_{\Omega}\left(\rho|\nabla v|^2+\frac{(v-1)^2}{4\rho}\right)dx \tag{1.1}$$

is discretized as $\rho\|\nabla v\|_2^2+\frac{\|v-1\|_2^2}{4\rho}$.

For each binary image, we compute its edge strength function based on the following diffusion model with Dirichlet boundary condition

$$v_i=\underset{v}{\operatorname{argmin}}\,\rho\|\nabla v\|_2^2+\frac{\|v-1\|_2^2}{4\rho},\quad v=0 \text{ on the boundary of the } i\text{th binary image.} \tag{1.2}$$

Fig. 1.1 Examples of silhouette images (*top row*) and their edge strength functions (*bottom row*)

Given a library

$$A = \begin{bmatrix} | & | & & | \\ v_1 & v_2 & \cdots & v_N \\ | & | & & | \end{bmatrix},$$

our goal is to learn a prior shape v such that

$$v = As + w + e = \sum_{i=1}^{N} s_i v_i + w + e,$$

where w is discrepancy and e is random Gaussian noise. Considering the sparsity of s and edge-like characteristic of w, we propose the following model:

$$\min_{s,w} \|\nabla w\|_1 + \beta \|s\|_0 \quad \text{subject to} \quad \|As + w - v\|_2 \le \epsilon,$$

where $\alpha, \beta > 0$ are parameters, and ϵ is the standard deviation of the error. By converting into the unconstrained minimization problem, the above model reads as

$$\min_{s,d} \frac{1}{2} \|As + w - v\|_2^2 + \alpha \|\nabla w\|_1 + \beta \|s\|_0 .$$

Since the ℓ_0 problem is NP-hard, we make a relaxation and solve the following ℓ_1 problem

$$\min_{s,d} \frac{1}{2} \|As + w - v\|_2^2 + \alpha \|\nabla w\|_1 + \beta \|s\|_1 .$$

The reason that we use the edge strength function for shape rather than any other informative indicator function (e.g., signed distance function) is that it has a natural connection to the segmentation problem via Mumford and Shah. Note that the edge strength function is nothing but the minimizer of (1.1). In the previous section we have explained that the edge strength function approaches to the edge indicator in the $L^2(\Omega)$-topology as $\rho \to 0$. Interestingly, as we increase ρ, edge localization weakens and v begins to act as a *morphology coder*: (1) v value at a domain point is a monotonically decaying function of the distance from the point to the domain boundary (the edge set); (2) the level curves of v are curvature dependent erosions of the domain boundary [3]. Thus, in our model, unlike many other shape prior based segmentation models, we do not distinguish inside and outside in the intermediate steps.

1.2.3 Proposed Segmentation Model

Given a reference image $g(x)$, we propose the following segmentation model:

$$\begin{aligned}\min_{u,v,s,d,h} & \frac{1}{2}\|u-g\|_2^2 + \frac{\alpha}{2}\|v\cdot\nabla u\|_2^2 + \frac{\rho}{2}\|\nabla v\|_2^2 + \frac{\|v-1\|_2^2}{8\rho} \\ & + \beta\|\nabla w\|_1 + \tau\|s\|_1 + W(h) \\ & \text{subject to} \quad As + w = v(h).\end{aligned}$$

where $\cdot$ represents point-wise product, $W(h)$ is a regularization term with respect to h. Note that to make variables consistent in the above model $\|\nabla u\|_2^2$ and $\|\nabla v\|_2^2$ are the discretized versions of $\int_\Omega |\nabla u|^2 dx$ and $\int_\Omega |\nabla v|^2 dx$. Typically $W(h)$ is set as $\|\nabla h\|_2^2$.

The associated Lagrangian function is

$$\begin{aligned}L(u,h,v,s,w,t) = & \frac{1}{2}\|u-g\|_2^2 + \frac{\alpha}{2}\|v\cdot\nabla u\|_2^2 + \frac{\rho}{2}\|\nabla v\|_2^2 + \frac{\|v-1\|_2^2}{8\rho} \\ & + \beta\|\nabla w\|_1 + \tau\|s\|_1 + W(h) + \frac{\gamma}{2}\|As + w - v(h) - t\|_2^2\end{aligned}$$

where t is the scaled Lagrange multiplier and γ is a positive parameter. Since v and h are related and inseparable, we cannot directly apply the ADMM to solve the above model. As such, we consider the following modified ADMM with approximate subproblems:

$$\left\{\begin{aligned} u^{k+1} &= \operatorname*{argmin}_u \frac{1}{2}\|u-g\|_2^2 + \frac{\alpha}{2}\left\|v^k\cdot\nabla u\right\|_2^2, \\ v^{k+1} &= \operatorname*{argmin}_v \frac{\alpha}{2}\left\|v\cdot\nabla u^{k+1}\right\|_2^2 + \frac{\rho}{2}\|\nabla v\|_2^2 + \frac{\|v-1\|_2^2}{8\rho} \\ &\quad + \frac{\gamma}{2}\left\|As^k + w^k - v(h^k) - t^k\right\|_2^2 \\ h^{k+1} &= \operatorname*{argmin}_h \frac{\gamma}{2}\left\|As^k + w^k - v^{k+1}(h) - t^k\right\|_2^2 + W(h) \\ s^{k+1} &= \operatorname*{argmin}_s \tau\|s\|_1 + \frac{\gamma}{2}\left\|As + w^k - v^{k+1}(h^{k+1}) - t^k\right\|_2^2 \\ w^{k+1} &= \operatorname*{argmin}_d \beta\|\nabla w\|_1 + \frac{\gamma}{2}\left\|As^{k+1} + w - v^{k+1}(h^{k+1}) - t^k\right\|_2^2 \\ t^{k+1} &= t^k + \gamma(v^{k+1}(h^{k+1}) - (As^{k+1} + w^{k+1})) \end{aligned}\right. \tag{1.3}$$

The u-subproblem can be solved by applying the negative gradient flow

$$\frac{du}{dt} = -2(u - g) + \alpha \operatorname{div}\big((v^k)^2 \nabla u\big). \tag{1.4}$$

Likewise, the v-subproblem can be solved iteratively. The h-subproblem turns out to be a registration problem. Moreover, the s-subproblem and the w-subproblem are Lasso problems which can be directly solved by applying ADMM [4, 5].

1.3 Experiments

In this section, we show two numerical experiments to validate our proposed method. By the assumption that the desired edge strength function v is a linear combination of atoms in the library A, the atoms have to be linearly independent which ensures the unique representation of v in the column space of A. In addition, to avoid the interruptions of background during the learning process, we also restrict the data fidelity term in the s-subproblem to the union of shape interiors associated with atoms, which can be done by introducing the corresponding mask. In all our experiments, the library consists of five independent atoms which are generated by applying the model (1.2) to five binary shapes (see Fig. 1.2). The parameters for both experiments are set as $\rho = 8$, $\alpha = \gamma = 1$, and $\beta = 10^{-2}$.

At the first experiment, we test an image where a star is partially occluded by the background rectangles. After running 13 iterations, the desired atom corresponding to the star shape is learned from the library. The input image, the obtained edge strength function v and the extracted boundary by thresholding v with 0.015 respectively are shown in Fig. 1.3. This example shows that the proposed algorithm is able to find a matching shape from the library. At the second experiment, we test an image where a star with missing parts is contaminated by uniformly distributed Gaussian noise with zero mean and standard deviation $\sigma = 0.8$. After 60 iterations, the desired edge strength function v is obtained with noise reduction. The extracted boundary by thresholding v with 0.04 is shown in Fig. 1.4. One can see that the proposed method has a potential to supplement the insufficiency of input data by

Fig. 1.2 Atoms in the library used in the experiments

Fig. 1.3 From left to right: the input image which has a star occluded partially by the background, the output edge strength function v, and the extracted boundary

Fig. 1.4 From left to right: the input image which has a star in a noisy background with missing parts, the output edge strength function v, and the extracted boundary

learning a prior shape from the library. The resultant edge strength images can be further processed to obtain sharp boundaries of the objects by thresholding or other more sophisticated algorithms.

1.4 Conclusion

Shape prior plays an important role in segmenting images with occlusive and missing information. In this paper, we used edge strength functions as atoms of a library and applied sparse optimization methods to automatically and adaptively search for a shape prior from the library to guide segmentation in a variational scheme. Numerical experiments show that the proposed approach has some potentials in segmenting images with missing information, random noise and structure noise.

Acknowledgements The authors would like to thank Luminita Vese from the Department of Mathematics at the University of California, Los Angeles for insightful discussions. The joint research is partially funded via US NIH 1R21EB016535-01 to W.G. and TUBITAK 112E208 to S.T.

References

1. Mumford, D., Shah, J.: Optimal approximations by piecewise smooth functions and associated variational problems. Commun. Pure Appl. Math. **42**(5), 577–685 (1989)
2. Ambrosio, L., Tortorelli, V.: On the approximation of functionals depending on jumps by elliptic functionals via Γ-convergence. Commun. Pure Appl. Math. **43**(8), 999–1036 (1990)
3. Tari, S., Shah, J., Pien, H.: A computationally efficient shape analysis via level sets. In: Proceedings of the Workshop on Mathematical Methods in Biomedical Image Analysis, San Francisco, pp. 234–243 (1996)
4. Gabay, D., Mercier B.: A dual algorithm for the solution of nonlinear variational problems via finite element approximations. Comput. Math. Appl. **2**, 17–40 (1976)
5. Glowinski, R., Marrocco, A.: Sur l'approximation par éléments finis et la résolution par pénalisation-dualité d'une classe de problèmes de dirichlet non linéaires. Revue Française d'Automatique, Informatique, Recherche Operationnelle, Serie Rouge (Analyse Numérique) **9**, 41–76 (1975)

Chapter 2
A Scalable Fluctuating Distance Field: An Application to Tumor Shape Analysis

R. Alp Guler, Andac Hamamci, and Gozde Unal

Abstract Tumor growth involves highly complicated processes and complex dynamics, which typically lead to deviation of tumor shape from a compact structure. In order to quantify the tumor shape variations in a follow-up scenario, a shape registration based on a scalable fluctuating shape field is described. In the earlier work of fluctuating distance fields (Tari and Genctav, J Math Imaging Vis 1–18, 2013; Tari, Fluctuating distance fields, parts, three-partite skeletons. In: Innovations for shape analysis. Springer, Berlin/New York, pp 439–466, 2013), the shape field consists of positive and negative values whose zero crossing separates the central and the peripheral volumes of a silhouette. We add a non-linear constraint upon the original fluctuating field idea in order to introduce a "fluctuation scale", which indicates an assumption about peripherality. This provides the induction of an hierarchy hypothesis onto the field. When fixed, the field becomes robust for scale changes for analysis of correspondence. We utilize the scalable fluctuating field first in segmentation of the protruded regions in a tumor, which are significant for the radiotherapy planning and assessment procedures. Furthermore, the unique information encoded in the shape field is utilized as an underlying shape representation for follow-up registration applications. The representation performance of the scalable field for a fixed 'fluctuation scale' is demonstrated in comparison to the conventional distance transform approach for the registration problem.

2.1 Introduction

Tumor growth modeling is extensively studied using theoretical and experimental approaches by a variety of disciplines. While majority of the current studies are focused on modeling microscopic phenomena, mathematical models that operate

R.A. Guler (✉) • A. Hamamci • G. Unal
Faculty of Engineering and Natural Sciences, Sabanci University, Istanbul, Turkey
e-mail: alpguler@sabanciuniv.edu; andachamamci@sabanciuniv.edu; gozdeunal@sabanciuniv.edu

K. Leonard, S. Tari (eds.), *Research in Shape Modeling*, Association for Women in Mathematics Series 1, DOI 10.1007/978-3-319-16348-2_2

at a macroscopic level are increasingly investigated through the analysis of clinical medical images [23]. Inhomogeneous and anisotropic tumor growth mechanisms lead to deviations of the tumor's shape characteristics from a compact structure and include protrusions. It is clear that extracting and quantifying the spatial information that irregular tumor shape parts carry would be a helpful macroscopic research tool for a better understanding of the dynamics of tumor growth.

As for clinical usage, the quantification and segmentation of the protruded and peripheral tumor regions could play an important role in radiosurgical applications. The goal of radiosurgery is to deliver a necrotic dose of radiation to the tumor while minimizing the amount of radiation to healthy brain tissues, especially to dose-sensitive tissues [39]. Series of beam configurations are determined as an optimization problem for treatment planning process such that beams will intersect to form a high dose at the tumor ROI. The rapid decrease at the edges of the radiation beam, which corresponds to the between 80 and 20 % isodose lines, is called the penumbra region and is generally located on the peripheral regions of the tumor [22]. A model that allows the distinguished analysis of the peripheral regions and segmentation of these parts that receive less radiation dose would not only be useful for isodose planning, but also for evaluating the success of the operation on protrusions and peripheral regions that are in close relation to critical anatomical structures. We propose an interactive method to distinguish protruded-peripheral parts using solely distance relations.

Segmentation or partitioning of shapes as boundary meshes is a problem of great interest for geometric modeling and computer graphics fields. The partitioning of the object represented by the mesh into meaningful parts, referred to as *part-type segmentation* by Shamir[33], is highly motivated by the study of human cognition [4, 18]. For an in-detail analysis of existing mesh segmentation methods we refer to [10, 33], along with recent successful approaches [16, 21] and a comparison of part-type segmentation techniques can be found in [1]. Distance functions described on the shape surfaces are commonly utilized for shape decomposition. There is a variety of surface metrics, e.g. geodesic [14], isophotic [24, 31], diffusion [11–13], volumetric part aware [25]. Though successful with a mesh representation, adaptation of these decomposition methods that use distance metrics to a volumetric representation would not be plausible. Additionally, partitioning the protrusions of tumors would require the abstraction of peripheral regions beforehand, else the association of partitioned boundary segments to the tumor volume would not be possible.

A sound approach for regional shape partitioning is utilizing the medial axis of symmetry, i.e. skeleton representation [6]. Partitioning shapes by associating regions with medial locus branches is very common and also successfully utilized in medical imaging [20, 30, 34, 35]. However, skeletal representations commonly suffer from certain instabilities. One of the instabilities is due to boundary perturbations, which are commonly addressed using smoothing or branch pruning approaches, which involve discarding branches that contribute little to the reconstruction of the shape [3, 7, 32]. For partitioning, choice of branches to prune would affect the resulting decomposition drastically considering the highly compact shapes of tumors, which also tend to inherit symmetries. Another kind of instability occurs in the regions near the junctions, which is mainly referred to as the ligature problem [2, 5]. A variety of methods have been proposed to cope with the ligature problem, including detecting transitional areas [28], a Bayesian formulation for estimating likely branches that would produce the shape [15] or disconnected skeleton approaches [9, 26, 38]. Additional to these inconsistencies, the association of branches with protrusions is not straightforward and even under slight deformation the abstraction of the centrality of the shape is not possible for fold-symmetry cases, which are highly possible for tumor shapes. Tari's model of Three-Partite-Skeleton, which arises from fluctuating distance fields [36] adresses this problem, which is highly motivating for the purpose of protrusion segmentation.

The fluctuating distance field [36, 37] contains both positive and negative values, and its zero crossing separates central and peripheral volumes. The maximum value of the field can be considered as a rough approximation of the center point for the shape in question, for instance the tumor, whereas the local minima correspond to rough approximations of center points for the protruded parts on the shape. The level curves encode the spatial relationships so explicitly that the separate protruded parts can be segmented even using a watershed segmentation without any additional processing. The extracted central region is compact and the peripheral region is always partitioned, unless it is a perfect annulus. In this model, no control exists over the ratio of region cardinality of positive field values to that of the negative field values. However, such a property can be an advantage in forming a shape field that respects a certain scale of central to peripheral regions of the shape. Particularly for shapes of tumorous structures, where boundaries between peripherality versus centrality is rather vague, variation of such a scale will introduce a flexibility in following shape analysis stages.

In this paper, we describe a scalable fluctuating distance field as a tumor description model. This model allows the user to interactively adjust the ratio of positive and negative domain sizes. The corresponding parameter can be set according to nature of the application. Thanks to this addition, a hierarchy of parts is not to be abstracted from the field as in [37]. Instead, fields that represent different hierarchical assumptions are formed, with the trade-off of losing linearity of the formulation. Details about the formulation and implementation of the shape field will be described in Sect. 2.2, where the fluctuation scale space that arises with the new parameter is introduced and exemplified on 2D shapes and 3D tumor volumes.

The constructed shape fields will be used for an alignment of baseline and follow-up tumor structures. In this registration problem, the distance transform is often used as a shape representation that describes the spatial relationships within the moving and fixed shapes [29]. The adjustment of the location of the zero-level set of the new distance field impairs the effect of scale changes to the resulting field for a fixed fluctuation scale, making the field a robust underlying shape representation for registration purposes. The registration process is described in Sect. 2.3 and experiments using both synthetic data and patient data are evaluated in Sect. 2.3.1, where the scalable fluctuating distance representation is compared to the conventional distance transform representation.

2.2 Scalable Fluctuating Distance Field

The concept of fluctuating distance fields, introduced by Tari [37], involves the exploitation of local and global spatial interactions to achieve a field that consists of both negative and positive values. The zero-level set partitions the shape domain into Ω^+ and Ω^-, which corresponds to the central region, a coarse and compact shape, and the peripheral region, which includes all the protrusions of the tumor, respectively. The ridge points on the surface yields the Three-Partite skeletons indicated. Our main motivation in using the fluctuating distance field is the information inherently coded in the resulting level curves at the peripheral regions, which will allow the explicit treatment to peripheral regions for further analysis. In this section we will describe our modification of this method, which will provide the required flexibility and interactivity for our purpose. We will follow by introducing the arising scale-space and illustrating segmented protruded parts using different fluctuation scales for 2D shapes and 3D tumor volumes.

The fluctuating distance field, ω: $\Omega \rightarrow \mathbb{R}$ is a real valued function on a discrete lattice, $\Omega \subset \mathbb{Z} \times \mathbb{Z} \times \mathbb{Z}$, with a neighborhood system, $\mathcal{N}$. ω is generated by the minimization of linear combinations of regional and boundary energies, which are described over the shape domain Ω, as a function of ω.

2.2.1 *Energy Terms*

The regional energy consists of local and global terms that function as spatial regularizers. Tari [37] proposed a global regional energy, which is the squared average over the domain, connecting all the nodes using a global mean constraint:

$$E_{Global}(\omega_{i,j,k}) = \frac{1}{|\Omega|} \sum_{(l,m,n)\in\Omega} \omega_{l,m,n}^2 \tag{2.1}$$

Differentiating $E_{Global}(\omega_{i,j,k})$ over Ω leads to the following expression:

$$\frac{\partial E_{Global}(\omega_{i,j,k})}{\partial(\omega_{i,j,k})} = \frac{2}{|\Omega|} \sum_{(l,m,n)\in\Omega} \omega_{l,m,n} \tag{2.2}$$

which would be minimized if ω is composed of all zeros or is a fluctuating function, where positive and negative values cancel each other.

The local regional energy functions as a smoothness term. We use the sum of squared differences between neighboring pixels in a six neighborhood system, $\mathcal{N}(i, j, k)$ to obtain the required spatial smoothness for the ω field:

$$E_{Local}(\omega_{i,j,k}) = \sum_{(l,m,n)\in\mathcal{N}(i,j,k)} (\omega_{l,m,n} - \omega_{i,j,k})^2 \tag{2.3}$$

Differentiating this energy w.r.t $\omega_{i,j,k}$ results in the following expression, where Ł corresponds to the seven-point discretization of the Laplacian operator:

$$\begin{aligned}\frac{\partial E_{Local}(\omega_{i,j,k})}{\partial(\omega_{i,j,k})} &= -2(\omega_{i+1,j,k} + \omega_{i-1,j,k} + \omega_{i,j+1,k} + \omega_{i,j-1,k} \\ &\quad + \omega_{i,j,k+1} + \omega_{i,j,k-1} - 6\omega_{i,j,k}) \\ &= -2Ł(\omega_{i,j,k})\end{aligned} \tag{2.4}$$

The boundary energy is defined for formulating the interactions along the level surfaces. The preservation of interactions between the nodes is imposed on the ω field using the usual distance transform as a bridge [37]. Thanks to this constraint, central regions of the shape, where the distance transform has larger values have much higher tendency to get positive ω values. The similarity to the distance transform function is formulated as follows:

$$E_{Bdry}(\omega_{i,j,k}) = (\omega_{i,j,k} - \mathcal{D}_{i,j,k})^2 \tag{2.5}$$

where $\mathcal{D}$ denotes the distance transform of the shape. The derivative of E_{Bdry} w.r.t $\omega_{i,j,k}$ is then given as follows:

$$\frac{\partial E_{Bdry}(\omega_{i,j,k})}{\partial(\omega_{i,j,k})} = 2(\omega_{i,j,k} - \mathcal{D}_{i,j,k}) \tag{2.6}$$

Minimization of the combination of these energies results in a ω field that has low expected value, thus fluctuating (2.2), locally smooth (2.4) and resembling the distance transform of the shape (2.6).

2.2.2 A Sign Constraint to Control Fluctuation Scale

The natural location of the zero-level curve under the given constraints often becomes too close to the tumor boundaries, turning out to be a disadvantage while estimating a deformation between two ω fields. In addition, the ability to control the location of the zero crossing turns the ω field to a robust feature for an interactive tool for segmenting the protrusions on the tumor. Therefore we describe an additional global constraint to adjust the position of the zero crossing. The term is constructed as a quadratic expression forcing the sum of the signs of all nodes to be close to a predetermined ratio of the domain size, $|\Omega|$:

$$E_{Sign}(\omega_{i,j,k}) = \sum_{(i,j,k)\in\Omega}\left(\left(\sum_{(l,m,n)\in\Omega} sign(\omega_{l,m,n})\right) - \eta|\Omega|\right)^2 \tag{2.7}$$

where $\eta \in [-1, 1]$ corresponds to the ratio of the intended sum of the signs of all ω points to the number of points in the shape domain $|\Omega|$. While minimizing (2.7), η is chosen as the desired ratio of:

$$\eta = \frac{\sum_{(l,m,n)\in\Omega} sign(\omega_{l,m,n})}{|\Omega|} = \frac{|\Omega^+| - |\Omega^-|}{|\Omega^+| + |\Omega^-|} \tag{2.8}$$

Differentiating this energy w.r.t. $\omega_{i,j,k}$ would give:

$$\frac{\partial E_{Sign}(\omega_{i,j,k})}{\partial(\omega_{i,j,k})} = 4\sum_{(i,j,k)\in\Omega}\left(\left(\sum_{(l,m,n)\in\Omega} sign(\omega_{l,m,n})\right) - \eta|\Omega|\right)\cdot\delta(\omega_{i,j,k}) \tag{2.9}$$

For the approximation of the signum function in a differentiable manner, we used a regularized Heaviside function, then the impulse function $\delta(z)$ was approximated as the derivative of $H(z)$:

$$sign(z) = 2H(z) - 1 \simeq \frac{2}{\pi}\arctan(\frac{z}{\epsilon}), \quad \delta(z) \simeq \frac{1}{\pi}(\frac{1}{1+(\frac{z}{\epsilon})^2})(\frac{1}{\epsilon}) \tag{2.10}$$

where ϵ determines the steepness of the smoothed step and the impulse functions.

2.2.2.1 Formulation

The computation of ω is achieved by calculating the steady state solution to the linear combinations of the energy derivatives, which are described above. The combination of the energies is presented in a continuous formulation as follows:

$$\iiint_\Omega (\omega_{x,y,z} - \mathfrak{D}_{x,y,z})^2 + \left(\frac{1}{|\Omega|}\iiint_\Omega \omega(\alpha,\beta,\theta)^2\, d\alpha d\beta d\theta\right) + (\nabla\omega(x,y,z))^2 + \dots$$
$$\dots\left(\left(\iiint_\Omega sign(\omega_{\alpha,\beta,\theta}) d\alpha d\beta d\theta\right) - \eta|\Omega|\right)^2 dx\, dy\, dz \tag{2.11}$$

The solution is obtained by applying the method of gradient descent in the following expression:

$$\frac{\partial \omega_{i,j,k}(\tau)}{\partial(\tau)} = -\frac{\partial(\beta_1 E_{Local}(\omega_{i,j,k}) + \beta_2 E_{Global}(\omega_{i,j,k}) + \gamma E_{Sign}(\omega_{i,j,k}) + \beta_3 E_{Bdry}(\omega_{i,j,k}))}{\partial \omega_{i,j,k}}$$

where β and γ values are Lagrange multipliers for the given energies. As natural choices, β_1, β_2, β_3 parameters can be interpreted as 1 [37]. γ is the only Lagrange multiplier that calibrates the relationship between the values of $E_{Bdry}(\omega_{i,j,k})$ and $E_{Sign}(\omega_{i,j,k})$. γ only affects convergence speed when it is within appropriate limits, that is not larger than the maximum value of the $\mathfrak{D}$. We choose it as a normalization to the E_{Sign} of the ω field with the desired size of $|\Omega^+|$ using roughly a spherical zero-level set assumption. The iterative scheme on ω is revealed after an artificial time discretization in τ:

$$\frac{\omega_{i,j,k}^{n+1} - \omega_{i,j,k}^{n}}{\Delta\tau} = \text{Ł}(\omega_{i,j,k}) - \frac{1}{|\Omega|}\sum_{(i,j,k)\in\Omega} \omega_{i,j,k}^{n} - \left(\frac{1}{|\Omega|}\omega_{i,j,k}^{n} - \mathfrak{D}_{i,j,k}\right)$$
$$-\gamma \sum_{(i,j,k)\in\Omega}\left(\sum_{(i,j,k)\in\Omega} sign(\omega_{i,j,k}^{n}) - \eta|\Omega|\right)\delta(\omega_{i,j,k}^{n}) \quad (2.12)$$

For the third term above, as ω is calculated up to a scale, a weight of $1/|\Omega|$ is used as a weighting between the $\mathfrak{D}$ and the ω field.

2.2.3 A Space of Fluctuation Scales

The effect of the parameter η of the E_{Sign} term is not only to change the location of the zero-level set. Its combination with the zero-mean constraint changes the encoding characteristics of the whole domain. For instance, positive η values force the negativity of the nodes that belong to Ω^- much more compared to $\eta = 0$ to satisfy the zero mean constraint. The reason is that there are less number of nodes that are negative, so those have to be more negative to satisfy the zero mean condition. The opposite goes for the negative η values. This causes a diversity in the characteristics of the fields as η changes. A separate normalization can be applied to the positive and negative parts of the fields, which diminishes this effect if not desired.

We depict the resulting fluctuation scale-space for a hand shape in Fig. 2.1a, where $\omega(\tilde{x}, \eta)$ is presented for $\tilde{x}$ on a vertical line on the hand shape domain and the surface plot for the zero-crossing contour as a function of η is presented in (b).

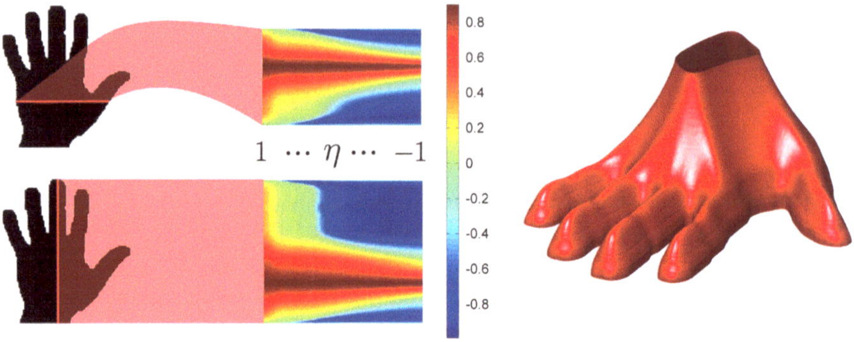

Fig. 2.1 (**a**) The normalized field $\omega(x = \tilde{x}, \eta)$, where $\tilde{x}$ is shown by horizontal (*top*) and vertical (*bottom*) *red lines*. Image obtained by sweeping η from 1 to -1. (**b**) Surface plot for $\omega(x, y, \eta) = 0$

Fig. 2.2 *From left to right:* Input shape, ω for $\eta > 0$, ω for $\eta = 0$, ω for $\eta < 0$

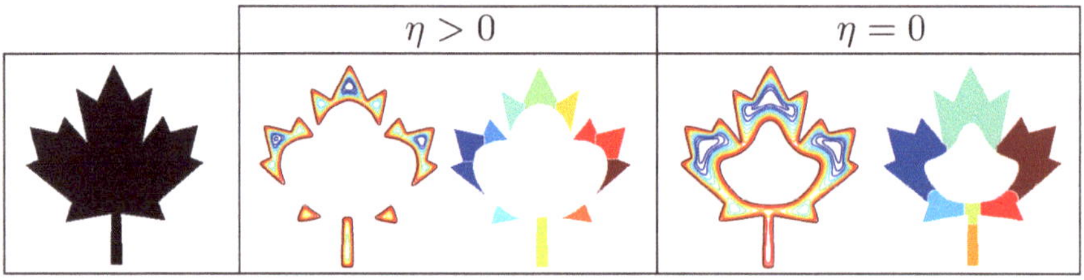

Fig. 2.3 Ω^- domain and watershed segmentation results for: *left* $\eta > 0$, *right*: $\eta = 0$

Notice that the zero-level set sweeps the whole domain smoothly from boundary to central regions, as the information regarding Ω is encoded for different scales of peripherality.

The computed field is shown for three different η values ($> 0, = 0, < 0$) for the symmetric shape silhouette in Fig. 2.2. Note that there are two levels of hierarchy in the peripheral regions of the shape, which can be seen as five different parts at a coarser level, later which are further differentiated into two separate parts. Varying the fluctuation scale parameter, one can capture those two levels of scale (coarser and finer) as can be observed in the resulting field with positive and negative η values, respectively.

A similar effect is achieved for the leaf silhouette in Fig. 2.3. Using a simple watershed segmentation [27], the resulting partitions reveal the three main leaves

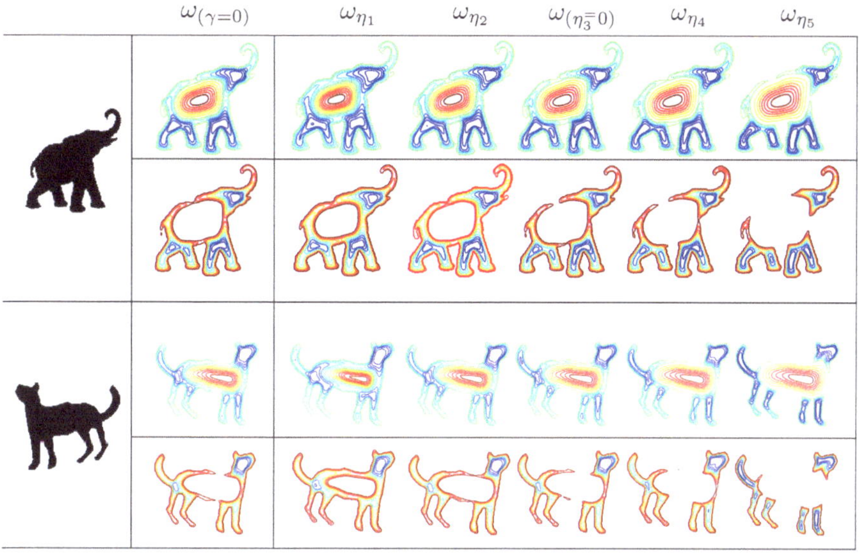

Fig. 2.4 The original ω field (*left*) [37], where the Lagrange multiplier γ is chosen as zero in Eq. 2.12 and five ω fields (*right*) calculated using increasing values for η, where $\eta_1 < \eta_2 < \eta_3 = 0$ and $\eta_5 > \eta_4 > \eta_3 = 0$. *Upper row* for both shapes is a contour plot of normalized ω and *bottom rows* depict ω for solely Ω^-

with $\eta = 0$, whereas the partitioning with the $\eta > 0$ field reveals the smaller protrusions on those three leaves. Here, the encoding of coarse to fine shape details nicely demonstrates the hierarchical aspect introduced into the fluctuating distance field.

We show the original w field and the scalable w field for various η values in Fig. 2.4 for an elephant and a cat silhouette. The first columns next to the silhouettes show the original field followed by the fields with increasing values of the fluctuation scale. The top picture is the whole w field, whereas the lower depicts only its Ω^- partition. Looking at the details at the legs of the fields more closely, for instance, the elephant's both front legs are merged in the original w field, as well as for the scalable field for smaller η values. When η is increased (e.g. see the rightmost field), the legs are separated, as can be observed in the Ω^--part of the field. This is because where the two legs are joined, there is a single local maximum with the original and low scale parameter fields, whereas there are two separate local maxima for each leg with the high-scale-parameter field. The same observation holds for the various shape fields over the cat. Note the rear-most leg of the cat and its tail which share a joint single maximum, whereas that extremum separates into two separate maxima for the tail and the rear leg towards the higher η-scale. Another point to remark over these experiments is the interesting feature of the low-η-fields when compared to the original w-field. Note the cat's front legs, and elephant's rear legs, which seem to have a separate maximum for each leg in the original shape field. However, the low η shape fields facilitate to peek at those same features first jointly then separately as the fluctuation scale varies from low to high. As these experiments demonstrate, the hierarchy over the shape is not built from the w-field as in [36], however, we modify the field itself to create the hierarchy that is sought for.

2.2.4 Interactive Tumor Protrusion Segmentation

The segmentation of the protruded tumor regions is achieved using the information in the negatively-valued regions of the ω field, which encapsulates local minima that depicts separate protrusions. The tumor should be segmented prior to the calculation of ω, for this purpose we use the Tumor-Cut method [17]. A contrast enhanced T1 MRI axial slice is depicted in Fig. 2.5, along with the ω field calculated on the tumor shape domain. Partitioning of the negatively-valued domain into protruded parts can be performed using the watershed transform [27] on the Ω^- field. The parts segmented from the resulting ω field can be observed Fig. 2.6 for a sample 3D tumor volume.

With the flexibility that E_{Sign} provides, the size of the positive compact part Ω^+ can be adjusted with user interference by medical experts or can be calculated

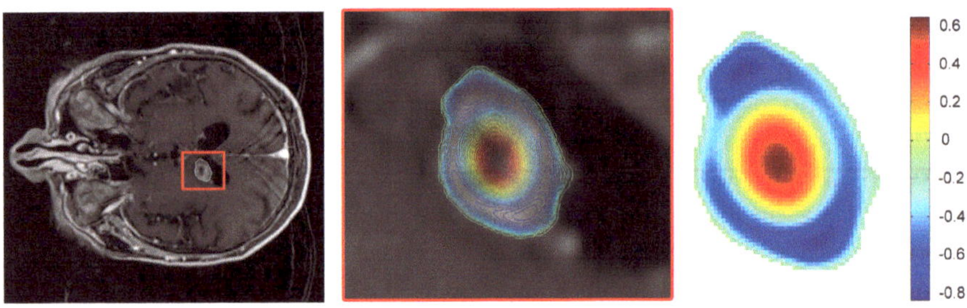

Fig. 2.5 *Left:* An axial slice of contrast enhanced T1 MRI of a patient with a tumor. *Middle:* ω field isocontours for the corresponding tumor slice. *Right:* ω field visualized

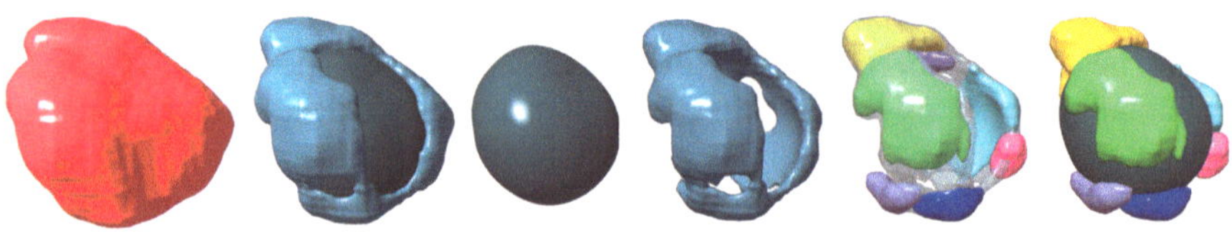

Fig. 2.6 *From left to right:* Tumor volume. Positive and negative parts of the proposed field. Positive part of the field. Negative part of the field. Segmented protrusions of the tumor enveloped in the negative part of the field. Segmented protrusions visualized with the positive part of the field

Fig. 2.7 Visualizations of positive(opaque) and negative(transparent) parts of the tumor field paired with corresponding segmentation results. The fluctuating distance field for each pair were generated using the corresponding η value

automatically by relaxing the η parameter until a predetermined hypothesis regarding the separated volumes are satisfied. The effect of η parameter on the resulting protruded parts is presented in Fig. 2.7.

2.3 Tumor Follow-Up Registration Using ω Fields

In order to obtain a valid and unbiased comparison between the performances of ω field and the conventional distance transform $\mathfrak{D}$ as underlying shape representations, we chose attributes that are essential in many of the registration algorithms that were proposed to calculate such deformations and combine them to end up with a basic yet powerful registration routine.

As linear data terms are not capable of performing well in case of large displacements, we used non-linear data terms and a coarse to fine warping approach which is a well studied combination in the area of optical flow estimation [8]. We follow the traditional model, formulated by means of an energy optimization problem, where deformation is calculated as a mapping between domains of shape fields ω_1 and ω_2. The displacement field $u \in \mathbb{R}^3 = (u_1, u_2, u_3)$ describes the deformation between the tumor and the follow-up shape domains: $u : \Omega_1 \in \mathbb{R}^3 \rightarrow \Omega_2 \in \mathbb{R}^3$. In the following: $x \in \mathbb{R}^3 = (x_1, x_2, x_3)$. The assumption of constancy of the underlying shape representation is formulated as:

$$\omega_1(x) - \omega_2(x + u) = 0$$

In addition to this data term, a regularization term based on the gradient of the deformation field is utilized. Following the original Horn and Schunck optical flow model [19], the combined functional F, where α is a parameter that controls the smoothness term:

$$F(u) = \int_{\Omega_1} (\omega_1(x) - \omega_2(x + u))^2 + \alpha^2(|\nabla u_1|^2 + |\nabla u_2|^2 + |\nabla u_3|^2)\mathrm{d}x \tag{2.13}$$

is minimized to yield the Euler-Lagrange equations, which are non-linear due to the $\omega_2(x + u)$ terms they contain. The first order Taylor expansions are used for those terms to obtain the linear system of three equations. First one of those three equations (for each coordinate) is written as:

$$(\omega_1(x) - \omega_2(x+u) - \nabla\omega_2(x+u)\,du)\,\omega_{2_{x_1}} + \alpha^2 div(\nabla u_1) = 0 \qquad (2.14)$$

where $\omega_{2_{x_i}}$ is the spatial derivative of ω_2 w.r.t x_i and $du \in \mathbb{R}^3$ describes an unknown update to the known variable u. In its solution, we adopted the warping scheme, introduced in [8], where the deformation field u is set to zero at the coarsest level and updated by $u^{n+1} = u^n + du$, as soon as du is computed at each finer scale using an inner loop of SOR iterations. $\omega_2(x+u)$ is computed at the beginning of each outer iteration by applying a warping process to $\omega_2(x)$ using the deformation field, u^n. The number of outer iterations depends on the downsampling factor. In order to achieve the full potential of the model, instead of the conventional 0.5 downsampling factor, we used a fixed value of 0.95 and a large number of outer iterations.

2.3.1 Registration Results and Discussion

Using both 2D synthetic shapes and real patient 3D tumor volumes, the performance of the ω field as an underlying shape representation for tumor follow-up registration is demonstrated against the distance transform, which is the conventional method to impose spatial shape relationships to the registration procedure.

Synthetic data results for pre-smoothed distance transform, pre-smoothed normalized distance trasform and fluctuating distance fields are respectively demonstrated in Fig. 2.8. On the top row for each of the experiments, the white and gray shapes denote the fixed and moving objects respectively, where the displacement vector field is demonstrated using arrows. Local volume change for each of the corresponding displacement field is generated using the determinant of the deformation gradient ($det(I + \nabla_x u)$) and presented below. The values of the determinant that are greater than 1 indicate a local expansion, whereas values less than 1 indicate a local contraction.

The distance transform, $\mathcal{D}$ is invariant to rotation and translation, but it is quite sensitive to scale changes[29]. Without a normalization, $\mathcal{D}$ representation can perform well for deformations without scale changes only, which certainly is not the case for tumor followup analysis. On the other hand, normalization causes an ambiguity in the information preserved in $\mathcal{D}$, leading to an estimation of the deformation field that does not fully describe the change between the shapes. However the ω field adopts less ambiguity, since the information is partitioned to separate parts, which leads to a robust estimation of the deformation. E_{Sign} constraint contributes highly to this robustness to scale change, for the ratio η will be the same in Ω_1 and Ω_2. Our experiments are highly coherent with this description. In Fig. 2.8, it is clear that $\mathcal{D}$ without a normalization fails to produce a smooth vector field. In addition while the local volume change in the deformation fields

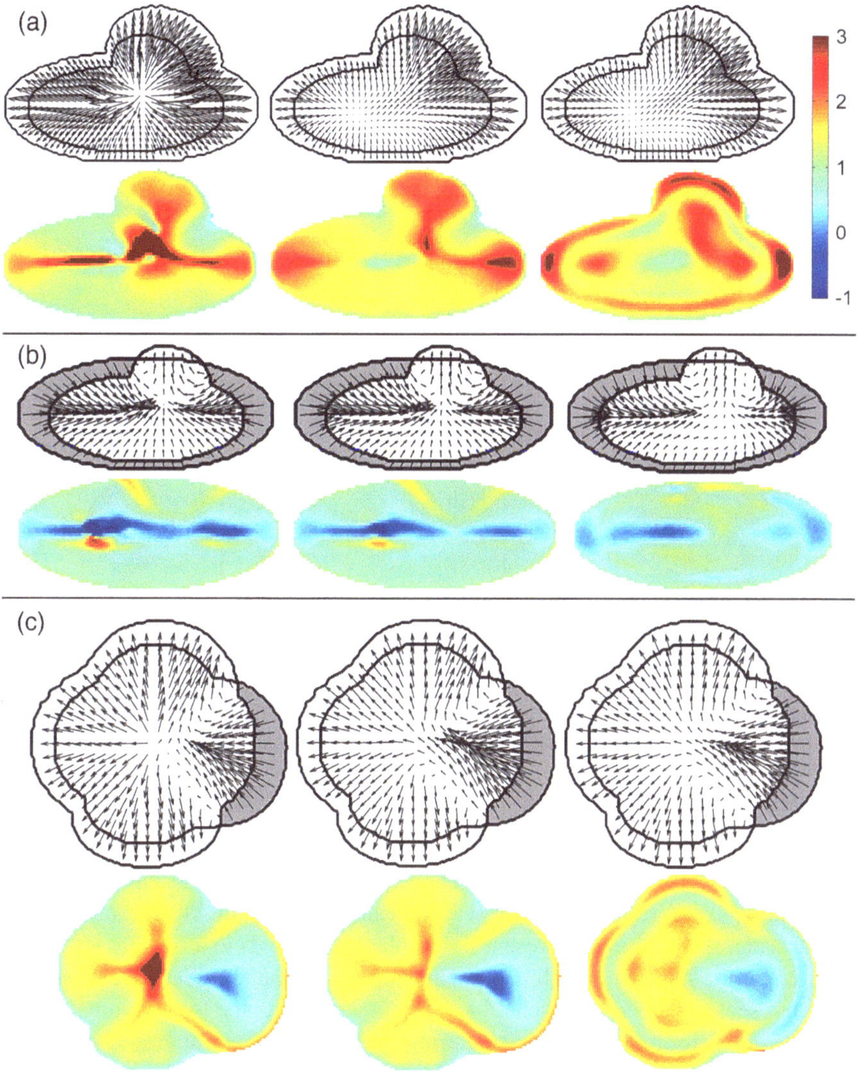

Fig. 2.8 (**a–c**): Visualizations of deformation field vectors and volume change pairs for registration of each synthetic shape couples, generated using *Left*: Distance transforms. *Middle*: Normalized High accuracy optical flow estimation based on a theory for warping.distance transforms. *Right*: Scalable fluctuating distance fields

estimated using ω is in accordance with the change in the shapes, the normalized $\mathcal{D}$ representation approaches fail to generate intuitive results. The expansions and contractions at the peripheral regions in Fig. 2.8 reveal the counter-intuitive nature of the displacement vectors generated using normalized $\mathcal{D}$.

Our experiments with patient data are demonstrated in Fig. 2.9, where the estimated 3D vector fields are visualized (on the left) for two pairs of tumor volumes

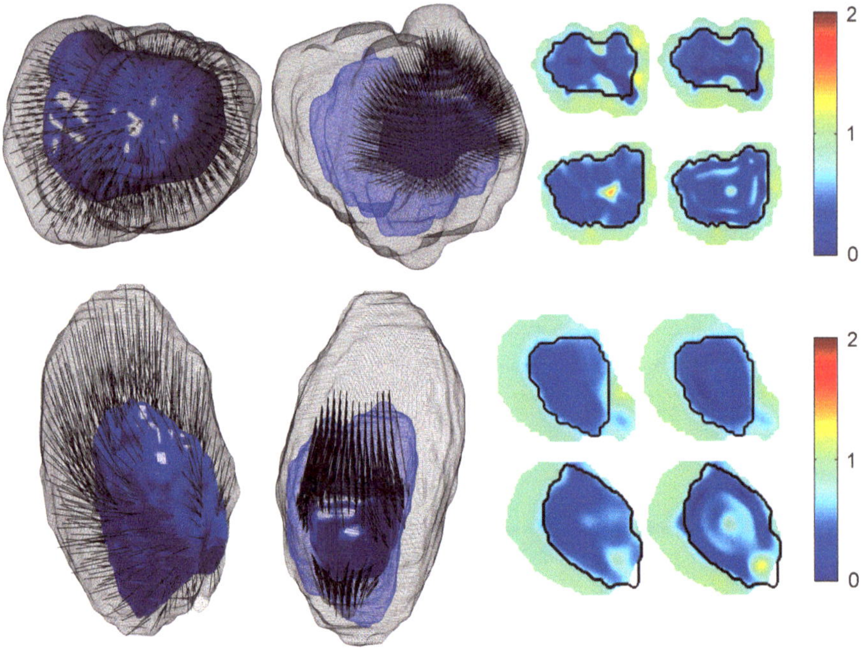

Fig. 2.9 For both parts of the figure: *Left:* Displacement field vectors from gray initial tumor to blue followup tumor. *Middle:* The displacement vectors to a specific segmented protrusion. *Right:* Local volume change maps in initial tumor domain for selected axial slices of the tumor shapes, the black contours denote the followup tumor. The maps on the left and right are generated from the deformation fields calculated using normalized $\mathcal{D}$ and ω fields respectively

on each row. Those pairs of tumor volumes are obtained after a segmentation on a pre-therapy and follow-up MRI scan and undergo a large change in terms of global scale. In addition, we present the displacement fields to a specific protrusion (Fig. 2.9 in the middle), which was segmented as described in Sect. 2.2.4. Various 2D cross sections are also depicted on the right along with the local volume change maps using deformation gradient determinants as explained above. Considering the large motion, necessity of regularization is quite larger in 3D tumor volumes compared to the phantom data in Fig. 2.8. For that reason the differences in the volume change maps are not as distinctive for the 3D volumes. But when these subtle changes are analyzed, they reveal the strength of the ω field in contrast to $\mathcal{D}$. The volume change maps of the second tumor shape in Fig. 2.9 is a convincing example: When the upper slice is analyzed it is clear that the deformation calculated using ω field (on the right) describes the compression smoother, yet on the bottom slice (right), it successfully represents the expansion while the distance transform approach is too smooth to describe an expansion. A similar robust behavior can be observed on the given local volume changes of the first tumor pair in Fig. 2.9.

2.4 Conclusion

Motivated from physical significance and clinical relevance in follow-up problems, we proposed a method to analyze the protruded and peripheral regions of tumor shapes. In order to introduce a parameter to control the fluctuation scales, we modified the fluctuating distance field [37] with an additional constraint on the ratio of sizes of the positive and negative domains which indicate central and peripheral shape regions respectively. This modification led to an interactive framework for segmenting the protrusions and partitioning tumorous structures, albeit the loss of the linearity of the original shape field model. The introduced nonlinear term due to its variable scale parameter, i.e. the "fluctuation scale", facilitates a hierarchical encoding of parts of the shape silhouette. By varying the fluctuation scale from low to high values, it is possible to observe the coarse to fine levels of hierarchy both in the field and its segmentations even by utilizing a very simple segmentation method.

The scalable shape field becomes a potentially powerful underlying shape representation for shape registration procedures, due to an increased robustness to scale changes without losing the information it inherits particularly in terms of the parts of a shape. For the registration application, the representation performance of the field was demonstrated in comparison to the conventional distance transform by observation of local volume changes in a tumor follow-up problem. Some counter-intuitive local changes were obtained by the latter, while the expected expansion and compression properties between pre-therapy and follow-up tumor volumes were provided by the deformation field estimated between the part-based shape fields.

References

1. Attene, M., Katz, S., Mortara, M., Patané, G., Spagnuolo, M., Tal, A.: Mesh segmentation-a comparative study. In: IEEE International Conference on Shape Modeling and Applications, SMI 2006, Matsushima, p. 7. IEEE (2006)
2. August, J., Siddiqi, K., Zucker, S.W.: Ligature instabilities in the perceptual organization of shape. In: IEEE Computer Society Conference on Computer Vision and Pattern Recognition, Fort Collins, vol. 2. IEEE (1999)
3. Bai, X., Latecki, L.J.: Path similarity skeleton graph matching. IEEE Trans. Pattern Anal. Mach. Intell. **30**(7), 1282–1292 (2008)
4. Biederman, I.: Recognition-by-components: a theory of human image understanding. Psychol. Rev. **94**, 115 (1987)
5. Blum, H., Nagel, R.N.: Shape description using weighted symmetric axis features. Pattern Recognit. **10**(3), 167–180 (1978)
6. Blum, H., et al.: A transformation for extracting new descriptors of shape. Models Percept. Speech Vis. Form **19**(5), 362–380 (1967)
7. Brady, M., Asada, H.: Smoothed local symmetries and their implementation. Int. J. Robot. Res. **3**(3), 36–61 (1984)
8. Brox, T., Bruhn, A., Papenberg, N., Weickert, J.: High accuracy optical flow estimation based on a theory for warping. In: Computer Vision-ECCV 2004, Prague, pp. 25–36 (2004)
9. Burbeck, C.A., Pizer, S.M.: Object representation by cores: identifying and representing primitive spatial regions. Vis. Res. **35**(13), 1917–1930 (1995)

10. Chen, X., Golovinskiy, A., Funkhouser, T.: A benchmark for 3d mesh segmentation. ACM Trans. Graph. (TOG) **28**, 73 (2009). ACM
11. Coifman, R.R., Lafon, S.: Diffusion maps. Appl. Comput. Harmon. Anal. **21**(1), 5–30 (2006)
12. Crane, K., Weischedel, C., Wardetzky, M.: Geodesics in heat: a new approach to computing distance based on heat flow. ACM Trans. Graph. (TOG) **32**(5), 152 (2013)
13. De Goes, F., Goldenstein, S., Velho, L.: A hierarchical segmentation of articulated bodies. Comput. Graph. Forum **27**, 1349–1356 (2008). Wiley Online Library
14. Do Carmo, M.P., Do Carmo, M.P.: Differential Geometry of Curves and Surfaces, vol. 2. Prentice-Hall, Englewood Cliffs (1976)
15. Feldman, J., Singh, M.: Bayesian estimation of the shape skeleton. Proc. Natl. Acad. Sci. **103**(47), 18,014–18,019 (2006)
16. Golovinskiy, A., Funkhouser, T.: Consistent segmentation of 3d models. Comput. Graph. **33**(3), 262–269 (2009)
17. Hamamci, A., Kucuk, N., Karaman, K., Engin, K., Unal, G.: Tumor-cut: segmentation of brain tumors on contrast enhanced mr images for radiosurgery applications. IEEE Trans. Med. Imaging **31**(3), 790–804 (2012)
18. Hoffman, D.D., Richards, W.A.: Parts of recognition. Cognition **18**, 65–96 (1984)
19. Horn, B., Schunck, B.: Determining optical flow. Artif. Intell. **17**, 185–203 (1981)
20. Joshi, S., Pizer, S., Fletcher, P.T., Yushkevich, P., Thall, A., Marron, J.: Multiscale deformable model segmentation and statistical shape analysis using medial descriptions. IEEE Trans. Med. Imaging **21**(5), 538–550 (2002)
21. Kalogerakis, E., Hertzmann, A., Singh, K.: Learning 3d mesh segmentation and labeling. ACM Trans. Graph. (TOG) **29**(4), 102 (2010)
22. Khan, F.M.: The Physics of Radiation Therapy. Lippincott Williams & Wilkins, Philadelphia/London (2009)
23. Konukoglu, E., Pennec, X., Clatz, O., Ayache, N.: Tumor growth modeling in oncological image analysis. In: Bankman, I. (ed.) Handbook of Medical Image Processing and Analysis – New edition, chap. 18, pp. 297–307. Burlington, San Diego, London (2008)
24. Lai, Y.K., Zhou, Q.Y., Hu, S.M., Wallner, J., Pottmann, D., et al.: Robust feature classification and editing. IEEE Trans. Vis. Comput. Graph. **13**(1), 34–45 (2007)
25. Liu, R., Zhang, H., Shamir, A., Cohen-Or, D.: A part-aware surface metric for shape analysis. Comput. Graph. Forum **28**, 397–406 (2009). Wiley Online Library
26. Macrini, D., Dickinson, S., Fleet, D., Siddiqi, K.: Bone graphs: medial shape parsing and abstraction. Comput. Vis. Image Underst. **115**(7), 1044–1061 (2011)
27. Meyer, F.: Topographic distance and watershed lines. Signal Process. **38**(1), 113–125 (1994)
28. Mi, X., DeCarlo, D.: Separating parts from 2d shapes using relatability. In: IEEE 11th International Conference on Computer Vision, ICCV 2007, Rio de Janeiro, pp. 1–8. IEEE (2007)
29. Paragios, N., Rousson, M., Ramesh, V.: Non-rigid registration using distance functions. Comput. Vis. Image Underst. **89**(2), 142–165 (2003)
30. Pizer, S.M., Fritsch, D.S., Yushkevich, P.A., Johnson, V.E., Chaney, E.L.: Segmentation, registration, and measurement of shape variation via image object shape. IEEE Trans. Med. Imaging **18**(10), 851–865 (1999)
31. Pottmann, H., Steiner, T., Hofer, M., Haider, C., Hanbury, A.: The isophotic metric and its application to feature sensitive morphology on surfaces. In: Computer Vision-ECCV 2004, Prague, pp. 18–23 (2004)
32. Shaked, D., Bruckstein, A.M.: Pruning medial axes. Comput. Vis. Image Underst. **69**(2), 156–169 (1998)
33. Shamir, A.: A survey on mesh segmentation techniques. Comput. Graph. Forum **27**, 1539–1556 (2008). Wiley Online Library
34. Styner, M., Gerig, G., Lieberman, J., Jones, D., Weinberger, D.: Statistical shape analysis of neuroanatomical structures based on medial models. Med. Image Anal. **7**(3), 207–220 (2003)
35. Styner, M., Lieberman, J.A., Pantazis, D., Gerig, G.: Boundary and medial shape analysis of the hippocampus in schizophrenia. Med. Image Anal. **8**(3), 197–203 (2004)

36. Tari, S.: Fluctuating distance fields, parts, three-partite skeletons. In: Innovations for Shape Analysis, pp. 439–466. Springer, Berlin/Heidelberg (2013)
37. Tari, S., Genctav, M.: From a non-local ambrosio-tortorelli phase field to a randomized part hierarchy tree. J. Math. Imaging Vis. **49**(1), 69–86. Springer (2014)
38. Tari, Z., Shah, J., Pien, H.: Extraction of shape skeletons from grayscale images. Comput. Vis. Image Underst. **66**(2), 133–146 (1997)
39. Tombropoulos, R., Schweikard, A., Latombe, J.C., Adler, J.: Treatment planning for image-guided robotic radiosurgery. In: Computer Vision, Virtual Reality and Robotics in Medicine, pp. 131–137. Springer, Berlin/Heidelberg (1995)

Chapter 3
Part-Aware Distance Fields for Easy Inbetweening in Arbitrary Dimensions

Sibel Tari

Abstract The motivation for this work is to explore a possible computer graphics application for a part aware distance field developed recently. Computing in-between shapes is chosen as a toy application. Rather than presenting a highly competitive scheme which continuously morphs one shape into another, our aim is to investigate whether in-betweens may be defined as ordinary averages once a proper shape representation (e.g. a part aware field) is established. The constructions are independent of the dimension of the space in which the shape is embedded as well as the number of shapes to be averaged.

3.1 Introduction

In-between shapes are of enormous interest in both theory of shape perception [5] and practice of computer graphics [3]. To compute in-betweens, the first shape needs to be continuously morphed into the second shape. In this work, we explore whether it is possible to express in-betweens by ordinary averages after choosing a proper field representation for shapes.

The experimented field representation is a combination of two different fields: The first field is activated on the interior whereas the second one on the exterior. The interior one is the recent ω field [10, 11], which fluctuates to yield intuitive parts. It is also possible to obtain a similar behaviour by switching eigenvectors [14]. The exterior field is the v-field [13], which implicitly encodes local convexity/concavity of the shape boundary.

A preview of the combined field f is given in Fig. 3.1. The input is a uniform grid 3D point cloud describing a cat shape in a volume. The first column depicts the entire field f. The next two columns depict its restriction to voxels of particular value, i.e., some level sets. Observe that the field implicitly codes parts and distances. Hence, we will call it a part-aware distance field. Given two fields

S. Tari (✉)
Department of Computer Engineering, Middle East Technical University, Ankara, Turkey
e-mail: stari@metu.edu.tr

K. Leonard, S. Tari (eds.), *Research in Shape Modeling*, Association for Women in Mathematics Series 1, DOI 10.1007/978-3-319-16348-2_3

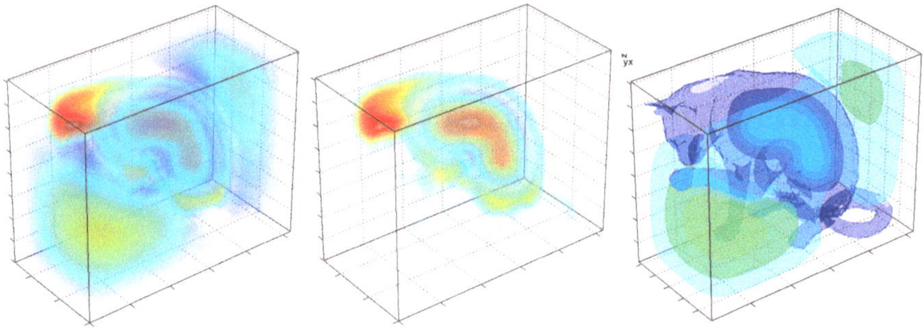

Fig. 3.1 The field as real valued mapping from $R^3 \rightarrow R$

computed from two different nD shapes, the idea is to define an *in-between* field as a weighted average of the two fields. The average is also treated as a part-aware field.

3.2 Computing the Field

The field f is computed as a solution to a linear system:

$$\mathbf{A}\,\mathbf{f} = \mathbf{b} \tag{3.1}$$

where $\mathbf{A}$ is an $N \times N$ matrix, where N is the number of points in a uniform discretisation of our bounded domain of interest, namely a subset of R^3 or R^2 containing the shape and part of its exterior; $\mathbf{f}$ and $\mathbf{b}$ are respectively the vector forms of the discretization of the function f and an external field b.

With a suitable ordering of the nodes as the ones belonging to the shape exterior, the shape boundary and the interior, $\mathbf{A}$ can be expressed in the following form:

$$\begin{bmatrix} [-\mathbf{L}+\alpha\mathbf{I}] & \mathbf{0} & \mathbf{0} \\ \mathbf{0} & [\gamma\mathbf{I}] & \mathbf{0} \\ \mathbf{0} & \mathbf{0} & [-\mathbf{L}+\alpha\mathbf{I}] \end{bmatrix} + \begin{bmatrix} \mathbf{0}\,\mathbf{0} & \mathbf{0} \\ \mathbf{0}\,\mathbf{0} & \mathbf{0} \\ \mathbf{0}\,\mathbf{0} & [\beta\mathbf{J}] \end{bmatrix} \tag{3.2}$$

where $\mathbf{L}$ is a discretisation of the Laplace operator, $\mathbf{J}$ stands for the matrix of ones, α and β are small scalars, say on the order of $\frac{1}{N}$; γ on the other hand is a large scalar.

The matrix $\mathbf{A}$ is block-diagonal. Relative to the linear operator $\mathbf{A}$, the function space R^N is a direct sum of three invariant subspaces: The second block serves to relax the Dirichlet condition on the shape boundary; turning the boundary condition (something that needs to be maintained) to an initial condition. Shape pairs are placed on a common domain, say rectangular. Homogeneous Dirichlet conditions are assumed on the domain boundary, whereas the ones on the shape boundary are *relaxed*. The first block on the left is positive definite matrix representing the modification of the matrix representation of the negative graph Laplace operator by some small multiple of the identity matrix, $-\mathbf{L} + \alpha\mathbf{I}$. If γ is selected very large, the second block approximates Dirichlet type boundary condition on the shape boundary, and the first two blocks becomes nothing but discretisation of the screened Poisson PDE. The third block of $\mathbf{A}$ is additively decomposed into two matrices where the one on the left $-\mathbf{L} + \alpha\mathbf{I}$ is the same with the first block, whereas the block on the right is a rank-1 matrix. The entire third block is a rank-1 modification of a symmetric positive definite matrix [2]. ($\mathbf{J}$ is rank-1.) Connection to rank-1 modification of the symmetric positive definite problems may help to interpret part-awareness of the field f; nevertheless, the system in (3.1) can be solved in numerous ways, including explicit solvers without needing to store large matrices.

For ease of implementation, we will solve two linear systems: one for the interior field and one for the exterior. For the interior field, the Dirichlet conditions are imposed on the shape boundary. For the exterior field, the Dirichlet conditions are imposed both on the shape and domain boundaries. This means that we solve for two fields as described in [11, 12] and [13] respectively.

In the next section, we further discuss the interior field.

3.2.1 The Interior Field: Spectral Perspective

Let us consider a field u

$$\mathbf{P}\mathbf{u} = \mathbf{1}$$

where $\mathbf{P}$ is a positive definite operator subject to homogeneous Dirichlet conditions, e.g. $-\mathbf{L}$ or $-\mathbf{L} + \alpha\mathbf{I}$. Spectral theorem says that the solution $\mathbf{u}$ can be expressed as a linear combination of the dot products of the right hand side $\mathbf{1}$ with each of the eigenfunctions. The weights are inversely proportional to the corresponding eigenvalues, making the flat mode the most dominant. The eigenfunctions of the Laplace (in Euclidean setting) or Laplace-Beltrami (in manifold setting) are very well studied; a great deal of shape analysis techniques are built on top of them, e.g., [9]. Interestingly, each mode, other than the first one, changes sign; thus, dividing the shape domain into nodal domains (not exactly parts in the perceptual sense). This naturally leads to an idea of obtaining a new operator by suppressing the flat mode. Let us consider changing the order of the first and last modes by switching the corresponding eigenvalues. Figure 3.2 depicts the fields constructed on cat shaped domains using this hypothetical operator. Observe how parts are revealed. This idea has been explored in [14].

The construction of the discussed hypothetical operator is expensive as it requires an explicit computation of the eigenfunctions. Such computations for volumes (rather than subsets of co-dimension 1) would be much more expensive. Computations get further complicated due to our need to relax boundary conditions.

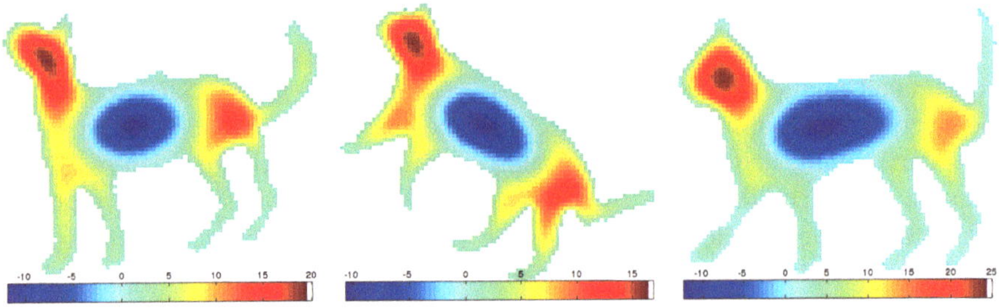

Fig. 3.2 A hypothetical part-aware field constructed on a variety of cat shaped domains using a hypothetical operator [14]. The operator is obtained by switching the first and the last modes of the Laplace operator in order to suppress the flat mode. Observe that a central body is cut out and peripheral parts are revealed

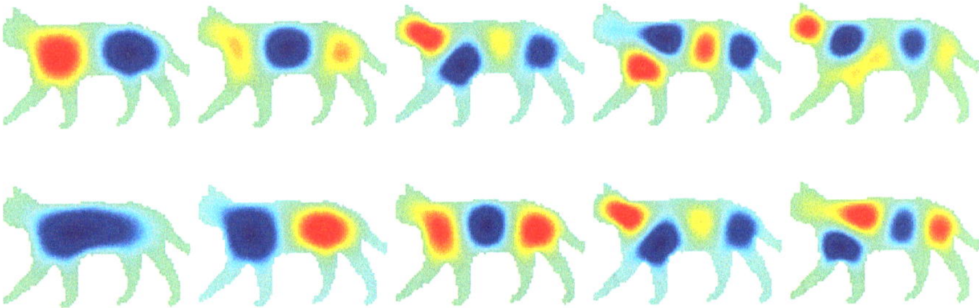

Fig. 3.3 Compare the modes of the third block of **A** (*the top row*) to those of the Laplace operator (*the bottom row*). Notice the resemblance of the jth mode to the $j+1$th mode of the Laplacian

In order to convey that our approach intuitively boils down to flat mode suppression, we depict the modes of the restriction of the **A** matrix to the shape interior (the third block). In the top row of Fig. 3.3, the first five modes are depicted. Even the first mode oscillates. Compare these modes to those of the Laplace operator depicted in the bottom row. Notice the resemblance of the jth mode to the $j+1$th mode of the Laplacian.

If the analysis is to be performed on a single shape, then the second block of **A** is be removed (imposing boundary conditions on the shape boundary) to arrive at a symmetric positive definite matrix, which is nothing but rank-1 modification of the Laplace operator. The general **A** matrix may not be positive definite. In our approach, positive definiteness becomes irrelevant because the field **f** is computed by directly solving a linear equation: (3.1). That is, spectral entities are not used.

Below, we give the remaining details necessary to compute the field **f**. That is, how to construct the right hand side. Solving (3.1) can be done in numerous ways.

3.2.2 *The Right Hand Side* b

In our implementation the right hand side function **b** is a concatenation of three vectors reflecting the three blocks of **A**.

$$\mathbf{b} = \begin{bmatrix} \mathbf{b_1} \\ \mathbf{b_2} \\ \mathbf{b_3} \end{bmatrix} \text{ s.t. } \mathbf{b_1} = \begin{bmatrix} b_1 \\ \vdots \\ b_{N_1} \end{bmatrix}; \mathbf{b_2} = \begin{bmatrix} b_{N_1+1} \\ \vdots \\ b_{N_2} \end{bmatrix}; \mathbf{b_3} = \begin{bmatrix} b_{N_2+1} \\ \vdots \\ b_{N_3} \end{bmatrix}$$

There are N_1 points in the exterior, $N_2 - N_1$ points on the shape boundary and $N_3 - N_2$ points on the interior. The middle group sets the *relaxed* Dirichlet conditions on locations indicating the bounding surface of the point cloud. This is only to allow averaging fields of different shapes. No parametric surface form is necessary. The first group is the constant vector **1**. The last group on the other hand is the discrete Euclidean distance function of the shape interior.

3.3 Experimental Results

We start by depicting the field $f : R^3 \rightarrow R$. A volumetric visualization with high transparency has already been given for one cat shape (Fig. 3.1). A noteworthy difference of the field from a signed distance function is that the new field changes sign twice: (1) to distinguish shape interior from shape exterior; (2) to distinguish central body from peripheral body. This causes a natural partitioning of the interior into meaningful entities. To provide more intuition, in Fig. 3.4, we depict iso-level surfaces that fall in certain ranges. We visualize corresponding sub volumes using high transparency.

We used another cat shape. In Fig. 3.5, we depict some individual iso-level surfaces. All of them are sampled from the interior. The blue surfaces are the ones where f takes on negative values. Samples are ordered in increasing magnitude. They code limbs and parts. As the level increases, only the most dominant parts survive. The red surfaces are the ones where f takes on positive values. They code central body. Samples are ordered in increasing magnitude.

In Figs. 3.6–3.9, we make parts explicit. Red opaque blob is the central part. In each case, it is naturally distinguished by the second sign change of f. Saddle points reveal other parts. To extract them, we simply call Matlab's watershed routine, which returns labeled groups. To convey 3D form, we used high transparency for the peripheral parts. In the pdf form, figures can be zoomed. One can observe how consistent and intuitive the parts are. Pay particular attention to the resemblance of the red opaque central parts across the horses and centaurs. This coarse central structure takes a different form for the gorilla (Fig. 3.9). We do not perform any pre- or post-processing on the field. Of course further clever organizations are possible

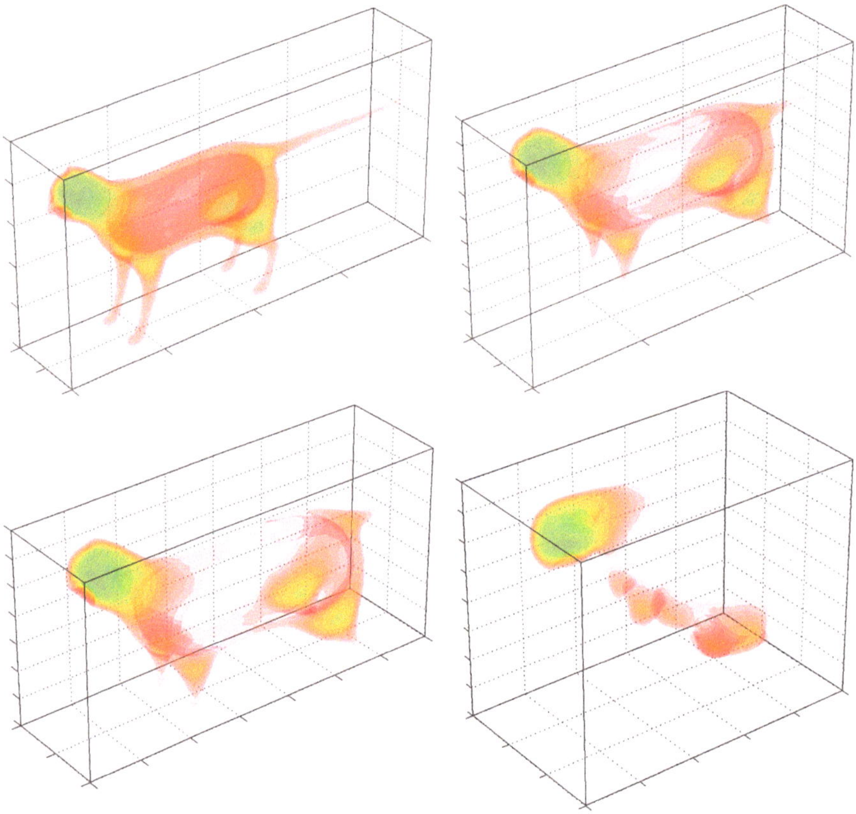

Fig. 3.4 Each figure depicts a range of iso-level surfaces. The respective volume is visualized using high transparency

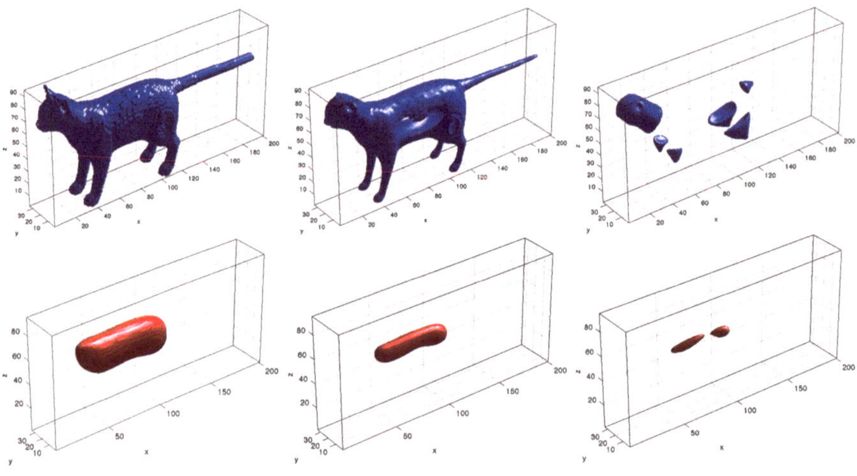

Fig. 3.5 Each figure depicts a single iso-level surface. All of the iso-level surfaces are sampled from the interior. On the *blue* ones, f takes on negative values. On the *red* ones, f takes on positive values

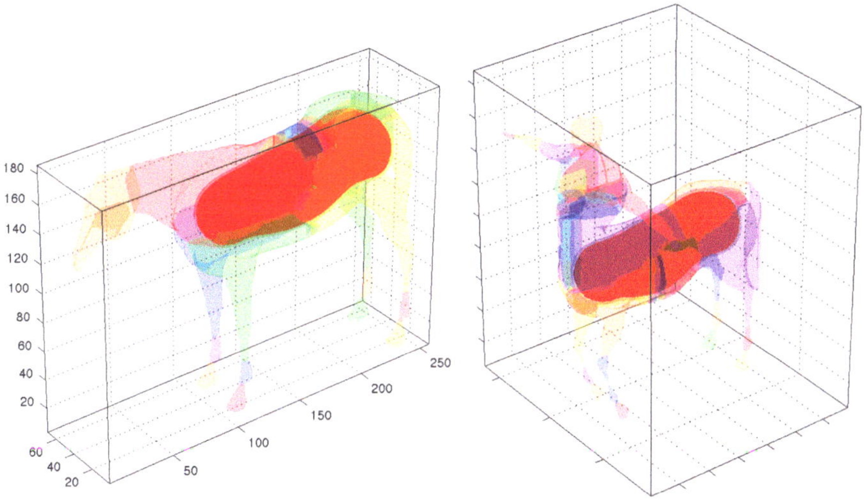

Fig. 3.6 Parts for a horse and a centaur

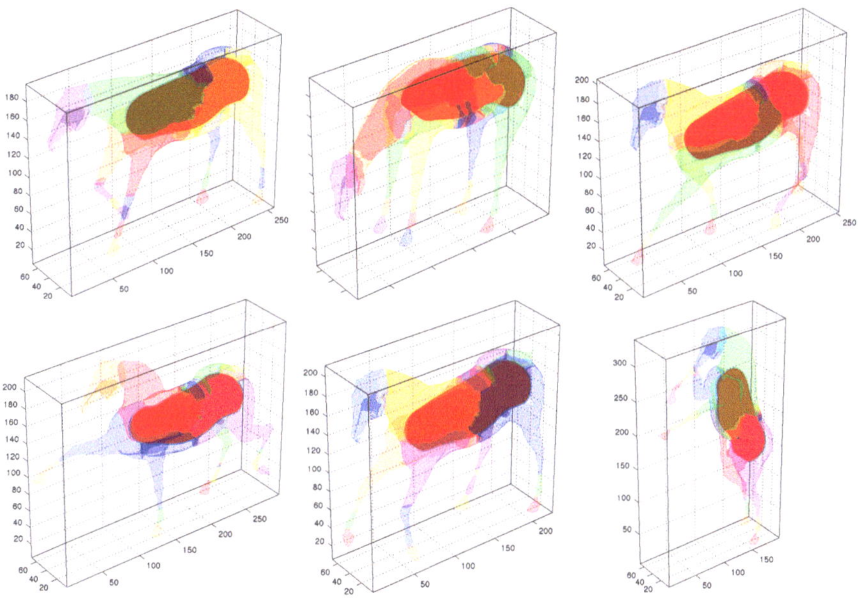

Fig. 3.7 Parts for more horses

as in [12]. Our purpose is to show results in their un-polished raw forms. Because we intend to remain implicit whenever possible, unless required otherwise in the context of a particular task.

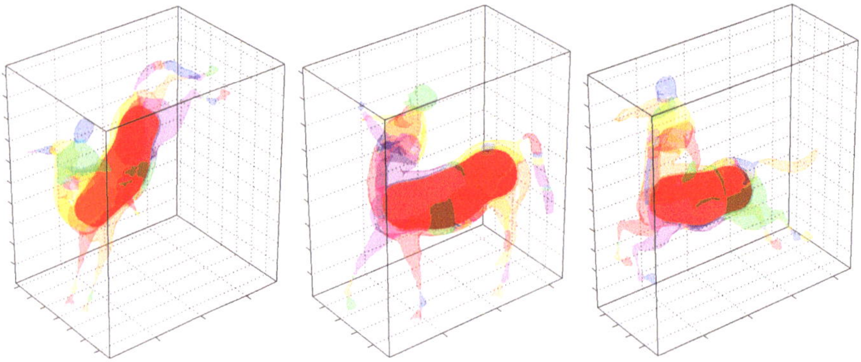

Fig. 3.8 Parts for more centaurs

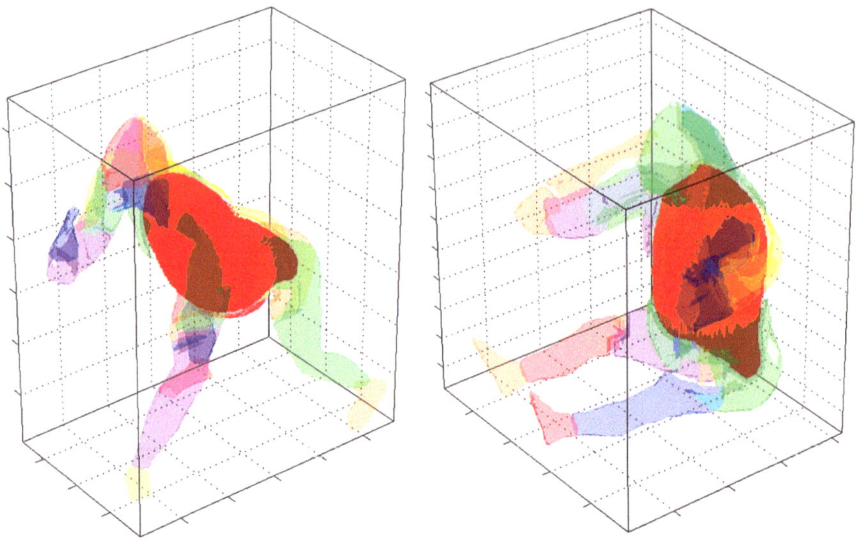

Fig. 3.9 Parts for gorillas

Our final three experiments are to illustrate algebraic operations on pairs of fields (Figs. 3.10–3.12). Due to linearity of our construction, it mathematically makes sense to compute a joint field of a (weighted) collection of shapes. What we particularly wonder is how much it makes sense to apply algebraic operations on collections (or pairs) of fields computed for individual shapes.

We start with an easy to interpret pair: a circle and a star of which centers coincide. Figure 3.10 depicts ten weighted averages, starting with a weight of $(1, 0)$ and ending with a weight of $(0, 1)$. The first and last figures are the individual fields for the circle and the star, followed by the parts extracted from the respective fields as splits via saddle points (Matlab's watershed command does this). The remaining eight figures are the parts extracted from average fields. No post- or pre-processing

Fig. 3.10 From a circle to a star, *in-between* shapes

is performed. We can observe the passage from a circle to a star. The initial averages are less star-like whereas the later ones are more star-like. The average shape in this case almost coincides with the union of the circle and the star. Of course, one can not form weighted unions.

When we consider, averaging a vertical ellipse with a horizontal one (tall ellipse and fat ellipse), the part structure of the mean field is similar to the part structure of a square: a shape equidistant from either of the ellipses. In Fig. 3.11, we average the fields of a dino and a lion. The input shapes are taken from [3]. As before, the first and last framed pairs are the fields and parts. Finally, Fig. 3.12 depicts some weighted averages of a sphere and a star-like composite of cylinders.

3.4 Some Discussions

From general paradigms point of view, the presented field best fits in the category of the so-called functionally-based implicit models. In the eighties, functionally-based implicit models were presented as then radical as well as promising alternatives to parametric and mesh-based representations [1, 8]. The basic idea was simple: instead of explicitly defining the model, namely a closed manifold of codimension

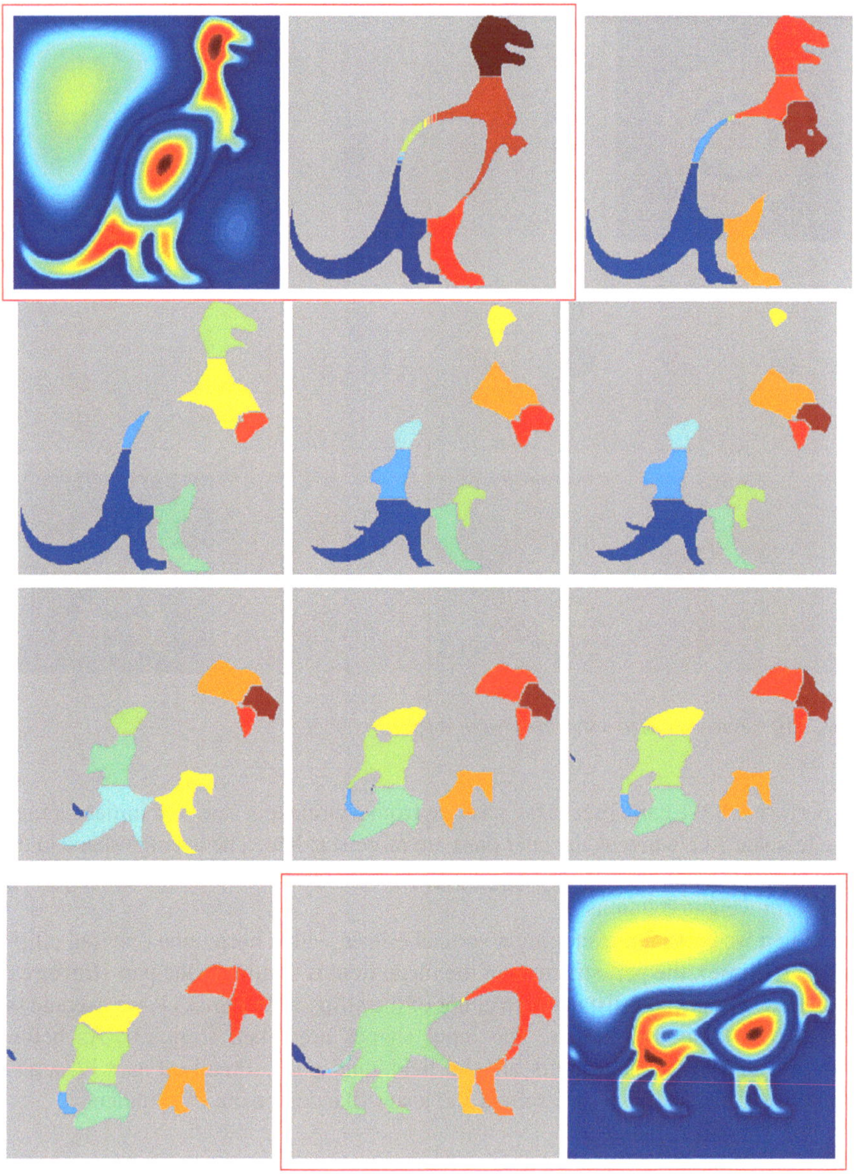

Fig. 3.11 From a dino to a lion, *in-between* shapes

one, a function $f : R^n \to R$ is constructed such that the three distinct values of the signum function at a point in the domain determines whether the point belongs to the interior, the boundary (the model), or the exterior. It is appropriate to say that f is an implicit model defining function.

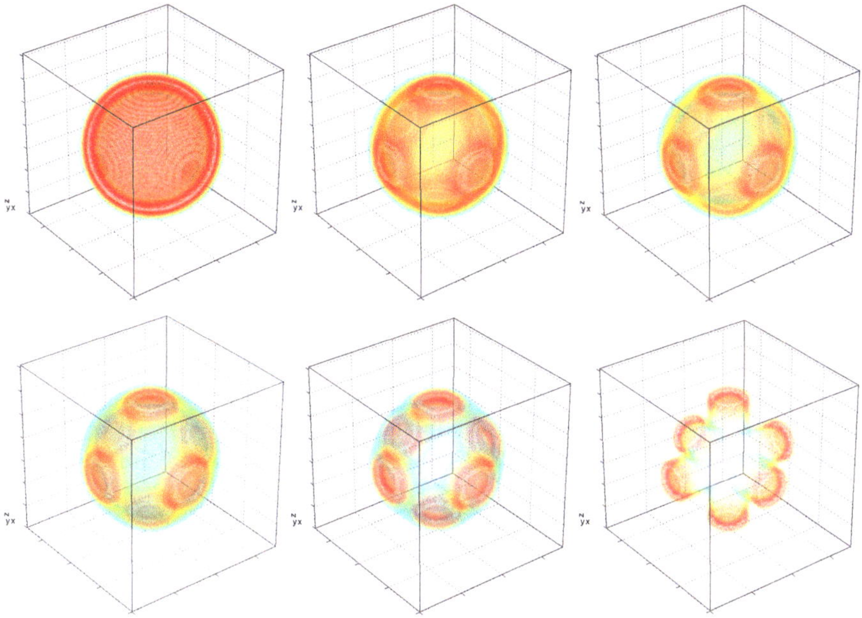

Fig. 3.12 From a sphere to a star made of cylinders, *in-betweens*

Early implicit model defining functions were typically either (1) polynomials of which roots indicate shape boundaries in the form of algebraic surfaces or (2) Euclidean distance functions. Later on, in mathematical imaging and shape analysis, Euclidean distance functions gave way to smooth distance fields [13]. There may be of course several different methods for constructing a smooth distance field, but no matter how the field is computed – be it by minimizing an energy or solving a screened Poisson or convolving – the key message is the same: take into account the local context in which a domain point is situated. Thus, smooth distance fields are more feature-aware where feature is boundary "curvature".

Does curvature-aware (or concavity-aware) imply part aware? Well, part perception is still an open problem. A classical paper by [4] relates part perception to local curvature minima. But, there are also psychophysical experiments which support that the boundary information is integrated relative to the interior; there are long-range, interior specific interactions occurring at early levels that explain formation of intermediate level representation [6]. There is even growing evidence that long-range interactions by themselves are not the only means and the global configuration can have an influence even on the perception of local features such as angles. That is where the part-aware distance proposal [11] comes in. The approach carries the concept of *model implicitation* one step further by capturing higher-level characterization in an implicit manner, too. The representation is computed from a single shape implied by a geometric point set. The resulting function representation allows efficient manipulation of shape combinations, supporting algebraic operations. In our modeling we remain implicit all the time without resorting to explicit features, assumed primitives, and explicitly designed geometric flows.

An alternative part-aware distance has been independently proposed in computer graphics [7]. Using the connection between a pair of surface points and a medial point in the interior, a so called *volumetric shape image* distance is built. The volumetric shape image distance is defined for each pair of surface mesh faces, and the distance between a pair of faces is zero if both faces map to the same medial point in the interior bounded by the surface; hence deemed to belong to the same part. Part-aware metrics suggest a promising direction of research with obvious uses. For example, in the context of partial shape matching, they may be used to limit the context of descriptors. The role of parts can not be denied in key graphics problems, e.g., co-analysis [15]. The part-aware surface metric [7], despite being an important idea, is subject to certain computational limitations; hindering its use and making it not as wide spread as diffusion distances defined over surfaces. Both the field f and ω eliminate these problems.

We explored the potential use of f field in a toy application of computing in-between shapes. We have demonstrated that the fields of individual shapes can be meaningfully averaged provided that an initial registration is performed. Our proposal neither coincides nor replaces existing *mean shape* understandings. We simply explore the representational limitations of a field based representation of shapes.

All of the constructions are independent of the dimension of the Euclidean space R^n in which the form is embedded.

Acknowledgements I thank M. Genctav for providing a faster code for the ω-field computation. The work is funded by TUBITAK 112E208.

References

1. Bloomenthal, J., Wyvill, B. (eds.): Introduction to Implicit Surfaces. Morgan Kaufmann Publishers, San Francisco (1997)
2. Bunch, J.R., Nielsen, C.P., Sorensen, D.C.: Rank-one modification of the symmetric eigenproblem. Numerische Mathematik **31**(1), 31–48 (1978)
3. Cohen-Or, D., Levin, D., Solomovici, A.: Contour blending using warp-guided distance field interpolation. In: IEEE Visualization, San Francisco, pp. 165–172 (1996)
4. Hofmann, D., Richards, W.: Parts of recognition. Cognition **18**, 65–96 (1984)
5. Klassen, E., Srivastava, A., Mio, W., Joshi, S.H.: Analysis of planar shapes using geodesic paths on shape spaces. IEEE Trans. Pattern Anal. Mach. Intell. **26**(3), 372–383 (2004)
6. Kovacs, I., Julesz, B.: Perceptual sensitivity maps within globally defined visual shapes. Nature **370**(6491), 644-646 (1994)
7. Liu, R., Zhang, H., Shamir, A., Cohen-Or, D.: A part-aware surface metric for shape analysis. Comput. Graph. Forum **28**(2), 397–406 (2009)

8. Pasko, A., Adzhiev, V.: Function-based shape modeling: mathematical framework and specialized language. In: Automated Deduction in Geometry, vol. 2930, pp. 132–160. Springer, Berlin/New York (2004)
9. Reuter, M.: Hierarchical shape segmentation and registration via topological features of Laplace-Beltrami eigenfunctions. Int. J. Comput. Vision **89**(2),287–308 (2010)
10. Tari, S.: Hierarchical shape decomposition via level sets. In: ISMM, Groningen pp. 215–225 (2009)
11. Tari, S.: Fluctuating distance fields, parts, three-partite skeletons. In: Innovations for Shape Analysis, pp. 439–466. Springer, Berlin/New York (2013)
12. Tari, S., Genctav, M.: From a non-local Ambrosio-Tortorelli phase field to a randomized part hierarchy tree. J. Math. Imaging Vis. **49**, 69–86 (2013)
13. Tari, S., Shah, J., Pien, H.: Extraction of shape skeletons from grayscale images. Comput. Vis. Image Underst. **66**(2), 133–146 (1997)
14. Tari, S., Burgeth, B., Tari, I.: Components of the shape revisited. In: Cognitive Shape Processing, American Association Artificial Intelligence Spring Symposium, Stanford (2010)
15. van Kaick, O., Xu, K., Zhang, H., Wang, Y., Sun, S., Shamir, A., Cohen-Or, D.: Co-hierarchical analysis of shape structures. ACM Trans. Graph. **32**(4), 1–10 (2013)

Chapter 4
A Biomechanical Model of Cortical Folding

Sarah Kim and Monica K. Hurdal

Abstract A principal characteristic of the geometry of the brain is its folding pattern which is composed of gyri (outward hills) and sulci (inward valleys). We present a preliminary two-dimensional biomechanical model of cortical folding that is implemented computationally using finite elements. This model uses mechanical properties such as stress, strain, and body forces, corresponding to axonal tension, to model the shape of the brain during early cortical development. Despite its simplicity, the proposed model can be used to demonstrate the plausibility of tension generating cortical folds, as has been suggested in Van Essen (Nature 385(6614):313–318, 1997). In addition, this model is used to investigate folding patterns on different domain sizes.

4.1 Introduction

The human cerebral cortex is a thin folded sheet of neural tissue forming the outermost layer of the cerebrum (brain). Also called the gray matter, the cerebral cortex is approximately 1–4.5 mm thick with an overall average thickness of approximately 2.5 mm [7, 9]. It is a six-layered structure [17] and surrounds the white matter (see Fig. 4.1a). The surface area of the adult human cortex is about 1,692 cm^2 [14]. The highly convoluted shape of the cortex enables the large surface area to be fitted within the relatively small size of the cortical volume which is about 458 cm^3 [14, 25].

Each human brain folding pattern is unique. The cortex begins to fold during the 26th week of gestation in humans (out of 40 gestational weeks) [5], forming gyri (outward hills) and sulci (inward valleys). The cerebrum is separated into right and left hemispheres by a deep sulcus called the longitudinal fissure. Other major sulci separate the two hemispheres into six major lobes having specialized functions.

S. Kim • M.K. Hurdal (✉)
Florida State University, Department of Mathematics, Tallahassee, FL 32306-4510, USA
e-mail: skim@math.fsu.edu; mhurdal@math.fsu.edu

K. Leonard, S. Tari (eds.), *Research in Shape Modeling*, Association for Women in Mathematics Series 1, DOI 10.1007/978-3-319-16348-2_4

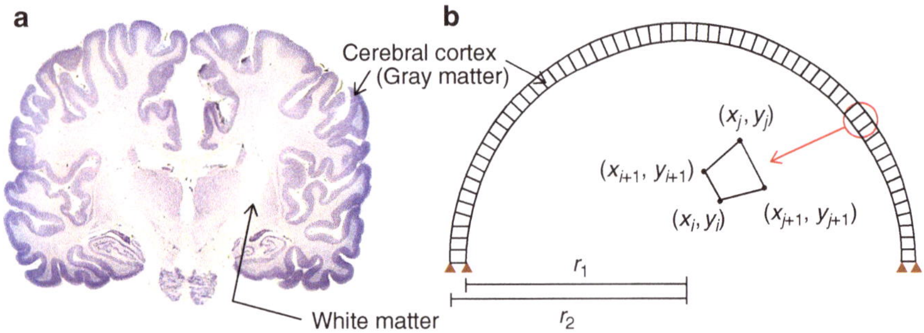

Fig. 4.1 Representation of the human brain. (**a**) Cross-section of the adult human brain taken perpendicular to the anterior-posterior axis (Adapted from [24]). (**b**) Computational model of the cerebral cortex. Quadrilateral elements, where (x_i, y_i) represents the Cartesian coordinates of the i-th node, are used. The radius of the gray matter, white matter, and cortical thickness are given by r_2, r_1, and $d = r_2 - r_1$ respectively

Primary sulci are most consistent in position across individuals while secondary sulci, which appear later in cortical development, are highly variable in position and appearance [9].

Understanding the underlying mechanisms in cortical folding will provide better strategies to treat neurological diseases. Many neurological diseases and dysfunction have abnormal cortical folding patterns. For example, Down's syndrome and lissencephaly patients have smoother brains as compared to healthy brains [6, 26]. On the other hand, polymicrogyria is a developmental malformation of the brain having an excessive number of small gyri on the cortex [2].

The mechanism producing cortical folding patterns is not fully understood. Several biological hypotheses and mathematical models have been proposed to elucidate the mechanisms involved in the development of cerebral cortical folding patterns. One of the leading hypotheses is tension-based morphogenesis proposed by Van Essen [25]. This hypothesis suggests that mechanical tension along axons in the white matter pull interconnected regions toward one another and it is the major inducing force for cortical convolution. This tension-based theory has been a foundation for formulating elastic mechanical models of cortical folding; see for example, [8, 15, 22].

Geng et al. [8] simulated the axonal tension forces at the lateral interior cortical surfaces using a stress-strain model, and the numerical results were consistent with the tension-based hypothesis. However, Xu et al. [28] have argued against the tension-based hypothesis by providing experimental data of developing ferret brains. According to their dissection data, Xu et al. claim that the location of subcortical white matter, where tension is present, is too deep to affect cortical folding. In addition, the dispersion of stress does not agree with Van Essen's hypothesis. Instead, Xu et al. support the differential cortical growth hypothesis of Richman et al. [17].

Richman et al. [17] have proposed that convolutional development of the brain is due to differential growth of the cortical layers and a significant differential in the elastic modulus of each cortical layer leads to cortical folding. They support this hypothesis by comparing two extreme cases of abnormal brains to a normal brain. In the normal brain, the outer cellular layers grow at a slightly faster rate than the inner layers. In microgyric cortex, the outer cortical layers grow at a slightly greater rate as compared to normal cortex while the inner cortical layers grow much slower (less than half) than in normal cortex. Yet, in lissencephalic cortex, there is no significant differential growth rate between cortical layers. Xu et al. [28] support this hypothesis with their computational models and experimental data.

Bayly et al. [3] also support differential growth, where tangential growth of the cortex is a mechanism driving cortical convolution. In contrast to Richman's differential growth hypothesis [17], Bayly et al. suggest cortical folding can be explained by differential growth even if the stiffness of the cortex and interior regions of the brain are similar. Other mechanical models of cortical folding pattern development, such as [15, 16, 22], suggest a major mechanism inducing cortical folding patterns is tangential cortical surface growth *without* the differential growth rate of the outer and inner cortical layers. Other authors suggest chemical factors, rather than mechanical factors, drive cortical folding [12, 19, 21].

In this paper, we propose a two-dimensional biomechanical model to elucidate mechanisms of cerebral cortical folding. The proposed model is based on the assumption that cortical convolutions are generated by axonal tension as suggested by [25] and supported by [15, 16, 22]. Since cortical growth and expansion is thought to occur preferentially along the tangential plane and tangential forces can be associated with axons in the white matter [25], we focus our simulations on tension due to tangential forces. By applying tension that pulls two regions on the cortical surface tangentially, we investigate whether gyri are formed by tension pulling interconnected regions toward one another. In addition, our model is able to simulate the effects of axonal tension in different directions while some previous models only considered axonal tension that pulls in the radial direction [22]. We investigate the effects of tension magnitude and direction by comparing folding due to tension pulling tangentially on a simulated cortical surface and tension pulling radially toward the inside of the cortex. Further, we simulate cortical folding on brains of different sizes to compare the degree of convolution between smaller brains and larger brains.

4.2 Biomechanical Model

Our proposed biomechanical model of cortical folding uses the theory of elasticity from classic solid mechanics. In this section, we describe the proposed model structure, define material properties, and describe the theory that we use. We use a finite element formulation to develop a numerical algorithm for the system of partial differential equations governed by the theory.

4.2.1 Brain Model Structure and Material Properties

We propose a two-dimensional semi-circular model geometry representing the initial shape of the cerebral cortex. The outermost layer of the cortex, corresponding to the gray matter, has radius r_2, and the boundary between the white matter and gray matter has radius r_1. We assume a uniform cortical thickness given by $d = r_2 - r_1$. For simplicity, we fix the location of the end points of the semi-circular model so they do not move in any direction during simulations. The semi-circular model is meshed into 200 quadrilateral elements (a schematic diagram is shown in Fig. 4.1b). Element nodes are labeled counter-clockwise beginning from the bottom right inner semi-circle.

The model structure is considered as an isotropic elastic material that has two independent constants in their stiffness and compliance matrices. For a material property matrix $[D]$ that makes a connection between the stresses and the strains, we assume the plane stress condition is applied as follows:

$$[D] = \frac{E}{1-\nu^2}\begin{bmatrix} 1 & \nu & 0 \\ \nu & 1 & 0 \\ 0 & 0 & \frac{1-\nu}{2} \end{bmatrix}, \tag{4.1}$$

where E is Young's modulus and ν is Poisson's ratio.

Brain tissue elasticity parameters are not readily available for a developing brain and we assume they are similar to those for a mature brain. We select elasticity parameters E and ν for the brain tissue obtained from experimental and statistical analysis in [18, 20, 23]. For our simulations, brain tissue elasticity parameters were the root-mean-square values obtained from the computational results of [18] and are $E = 9{,}210.87\,\text{Pa}$ and $\nu = 0.458344$. In addition, we use the data in [7, 16] for cortical thickness and the radius of the human brain during gestation. All length units were converted to inches and the unit of Young's modulus was converted from pascals (Pa) to psi.

4.2.2 Elasticity Theory and Finite Element Formulation

To explore several hypotheses regarding cortical folding, we assume that the behavior of the given material is regulated and deformed by the theory of elasticity. The governing coupled partial differential equations are vector equations with the displacements $u(x, y)$ and $v(x, y)$ in the x and y directions, respectively, as dependent variables [4].

We consider a static state, and equations of equilibrium can be stated as

$$\begin{aligned}\frac{\partial \sigma_x}{\partial x} + \frac{\partial \tau_{xy}}{\partial y} + f_x = 0 \, , \\ \frac{\partial \tau_{xy}}{\partial x} + \frac{\partial \sigma_y}{\partial y} + f_y = 0 \, ,\end{aligned} \tag{4.2}$$

where σ_x, σ_y, and τ_{xy} are stresses in the x and y directions, and the shear stress, respectively, and f_x and f_y are body forces per unit area along the x- and y-axes and which are assumed to be positive when acted along the positive axes. The body forces represent tension along axons when we simulate cortical folding with the proposed model.

By following the definition of strain which is used to measure changes in lengths or in angles, the relationship between strains and displacements can be stated as

$$\begin{aligned}\epsilon_x &= \frac{\partial u}{\partial x} \, , \\ \epsilon_y &= \frac{\partial v}{\partial y} \, , \\ \gamma_{xy} &= \frac{\partial u}{\partial y} + \frac{\partial v}{\partial x} \, ,\end{aligned} \tag{4.3}$$

where ϵ_x and ϵ_y are the extensional strains in the x and y directions, respectively, and γ_{xy} is the shear strain.

Applying the material property matrix $[D]$ in (4.1) gives the linear constitutive equations

$$\{\boldsymbol{\sigma}\} = [D]\{\boldsymbol{\epsilon}\} \, , \tag{4.4}$$

where $\{\boldsymbol{\sigma}\} = \{\sigma_x, \sigma_y, \tau_{xy}\}'$ and $\{\boldsymbol{\epsilon}\} = \{\epsilon_x, \epsilon_y, \gamma_{xy}\}'$. The set of equations in (4.4) provided by the stress-strain relations together with (4.2) and (4.3) completes the establishment of eight equations with eight unknowns: $u, v, \epsilon_x, \epsilon_y, \gamma_{xy}, \sigma_x, \sigma_y$, and τ_{xy}.

In order to develop a finite element formulation for (4.2)–(4.4), we used a weighted residual method, in particular Galerkin's method [13]. We discretized the domain using linear quadrilateral elements (see Fig. 4.1b). Each node has two degrees of freedom, and both displacements are interpolated using shape functions (see the Appendix). As a result, the element stiffness matrix $[K^e]$ for elasticity is expressed as

$$[K^e] = \int_{\Omega^e} [B]'[D][B] d\Omega \, ,$$

where $[D]$ is the material property matrix in (4.1) and $[B]$ is a 3×8 matrix evaluated by the linear shape functions on the four nodes of each quadrilateral element.

During development, cortical folding changes are gradual and occur over a long period of time. Linear constitutive equations are a reasonable choice as the brain is undergoing no large deformations. Furthermore, Wittek et al. [27] show that the choice of constitutive model of brain tissue has almost no influence on the computed deformation field and suggest that one can use the simplest elastic linear model with any reasonable value of Young's modulus and Poisson's ratio.

4.3 Results and Discussion

In this section, we describe numerical results of simulations using our proposed model. We investigate the effect of tensions due to tangential (circumferential) forces pulling interconnected regions. Next, we compare the effects of tangential forces and radial forces. In addition, we simulate effects of the mechanical forces on cortical folding on different domain sizes. In the results that follow, we represent the initial geometric configuration of the model cortex as thin black curves and the resulting deformed configuration is shown as dashed red curves. Arrows indicate the direction in which forces are applied. For visualization purposes the arrows are shown slightly interior of the model rather than on nodes where the forces are applied.

4.3.1 Tension Pulling Interconnected Regions

Van Essen [25] proposed that the principal driving force for cortical folding is mechanical tension along axons in the white matter. More specifically, he suggested that outward folds or hills (called gyri) are generated by tension pulling interconnected regions toward each other; on the other hand, inward folds or valleys (called sulci), are formed between each pair of outward folds [25]. This theory is persuasive since it explains compact wiring in the central nervous system. The existence of tension along axons in sub-cortical white matter has been proven through experimental studies but it has been argued that this tension is too deep to affect cortical folding [28].

We use our mechanical model to simulate and investigate the direction of folding due to pulling tension in the tangential direction. We use $r_2 = 4.0464$ cm which is the radius of the human brain at a gestational age of 28 weeks [1, 16]. Cortical thickness is taken to be 2.5 mm which is the average thickness of the human cerebral cortex reported by [7]. Tangential forces that represent tension along axons are loaded at ten nodes and each force vector has magnitude of 0.01 lbf (pound force). Each pair of tangential forces pulls toward one another (see arrows in Fig. 4.2a, b).

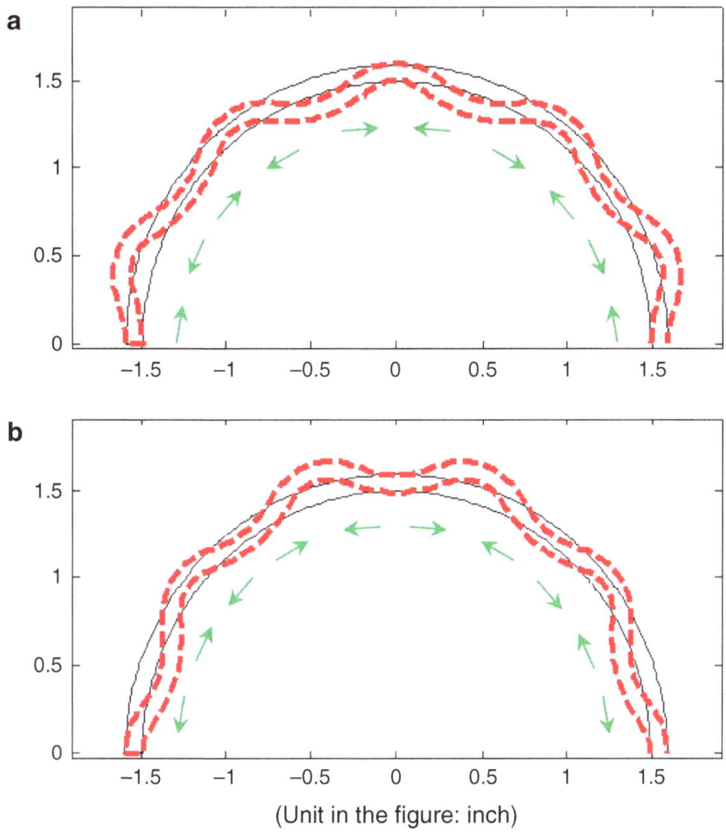

Fig. 4.2 Tension pulling interconnected regions. Model parameters are Young's modulus $E = 9{,}210.87\,\text{Pa}$, Poisson's ratio $\nu = 0.458344$, gray matter radius $r_2 = 4.0464\,\text{cm}$ (28 weeks gestational age), cortical thickness $d = 2.5\,\text{mm}$. Tangential forces of 0.01 lbf are applied at 10 nodes (*arrows* in (**a**) and (**b**)). The initial representation of the cortex (*black curves*) folds because each pair of tangential forces causes the cortex to pull together (*dashed red curves*)

These numerical results are consistent with the hypothesis of Van Essen. Each pair of pulling tension forces generates outward folds. Not surprisingly, inward folds are generated between the outward folds. In addition, when we simulated the pulling tension by using our model on a two-dimensional planar domain that has no curvature, the results of folding directions were also consistent with the tension-based theory of Van Essen.

Increasing the magnitude of the tangential forces causes the resulting folds to be deeper (see Fig. 4.3). Since our model uses realistic elastic parameters for the brain, it can be used to predict the magnitude of axonal tension necessary to generate folds of particular depths. For example, Fig. 4.3 shows that increasing the magnitude of the pair of tangential forces along the mid-line of the cortex representation results in a deep sulcus resembling the longitudinal fissure of the brain. By measuring specific

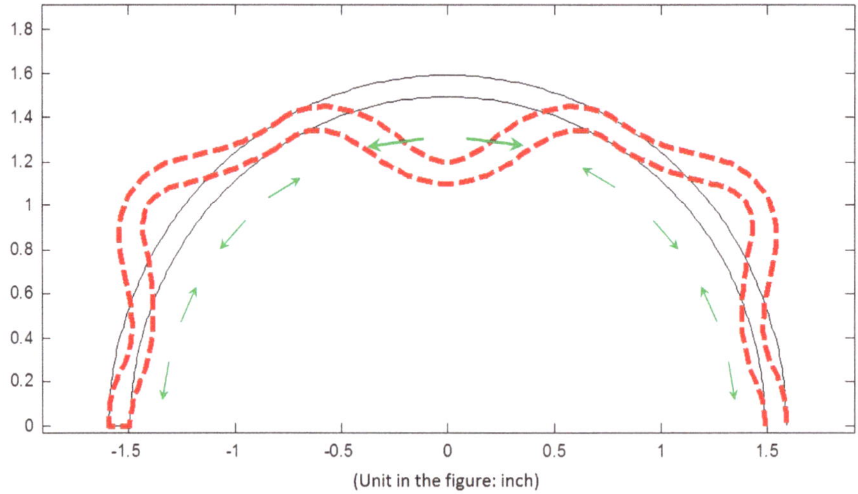

Fig. 4.3 Increasing force magnitude. Tangential forces are applied at 10 nodes (*arrows*). The pair of tangential forces that pull apart at two nodes near the center (mid-line) are 1.3 times the magnitude (*longer arrows*) of the other applied tangential forces. The cortical convolutions (*dashed red curves*) result in a deep sulcus, similar to the longitudinal fissure of the brain. Other parameters are as in Fig. 4.2b

sulcal depths from magnetic resonance imaging (MRI) data, our model can be used to estimate the magnitudes of axonal tension required to drive cortical convolution for those sulci. As measurements of axonal tension are currently unknown [8], it would be interesting to test these types of model predictions against experimental data.

4.3.2 Tangential Versus Radial Tension Direction

We investigate how tension direction along axons affects cortical folding. Tension in tangential directions on the surface and in radial directions (toward the center) are simulated by adjusting the body force components, f_x and f_y. Figure 4.4 shows the deformed configuration after radial forces are applied at (a) 4 nodes, and (b) 5 nodes, where the magnitude of each force vector is again 0.01 lbf (see arrows in Fig. 4.4a, b). For comparison, the results of applying tangential forces to 10 nodes where the magnitude of each force vector is 0.01 lbf (see Fig. 4.2a, b) is also shown in Fig. 4.4a, b. Other parameters, such as E, ν, r_2, and d, are the same as in Fig. 4.2.

Observe that the height of each fold is affected more by the radial forces than the tangential forces of the same magnitude. When radial forces are applied, the folds are deeper and the transition from a gyrus to a sulcus has increased curvature (i.e. shaper) even though fewer radial force vectors are applied. In addition,

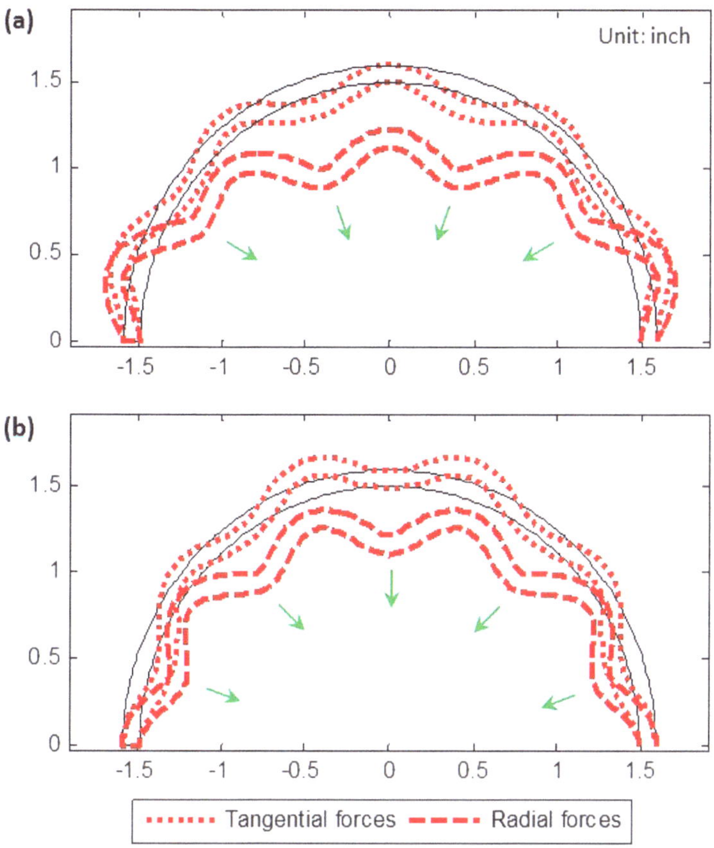

Fig. 4.4 Radial forces versus tangential forces. Inward radial forces of 0.01 lbf are applied at **(a)** 4 nodes and **(b)** 5 nodes (see *arrows*). Radial forces result in sulci that are deeper and more highly curved (*dashed red curves*) than when tangential forces are applied. For comparison, the effects of applying tangential forces of 0.01 lbf at 10 nodes that pull together in pairs are also shown (*dotted red curves*; see also Fig. 4.2). All other parameters are the same as in Fig. 4.2

the radial forces also affect the overall height of the model. The overall deformed configuration is "flatter" when inward radial forces are applied; in comparison, applying tangential forces causes the deformed shape to more closely resemble the height of the initial model configuration.

These simulation results indicate that increased tension along axons pulling radially or tangentially may account for cortical regions where sulci occur. However, radial forces account for deeper and sharper sulci as compared to tangential forces of similar magnitudes. If one were to assume that tension along axons is similar in magnitude regardless of direction of pull, then we hypothesize that tension due to tangential forces plays an increased role in the smoother cortical folding patterns of lissencephalic brains rather than tension pulling radially. Experimental data is needed to support this hypothesis.

4.3.3 Effects Due to Domain Size

Various studies investigate the degree to which cortical folding of the brain is affected by initial cortical shape, cortical volume, surface area, cortical thickness, and absolute mean curvature in sulcal regions [3, 10, 11]. In order to consider the effects of brain size and cortical thickness on cortical folding, we vary values of r_2 and d to affect gray matter radius (and hence brain size) and cortical thickness, respectively.

Simulation results in Fig. 4.5 change the value of r_2 while cortical thickness (d), elastic parameters (E and ν), and applied forces are the same as those used to simulate tension pulling interconnected regions in Fig. 4.2b. The values of r_2 increase in value in and represent the radius of the human brain taken at 11, 21, 28, 30, and 37 weeks gestational age [1, 16]. The simulation results show that the degree of folding is proportional to the size of the brain—a smaller brain has a smoother cortical sheet while a larger brain has a more convoluted cortex.

Simulation results in Fig. 4.6b–e use a variety of cortical thickness values (d) reported in [7]. As the cortex becomes thinner, the degree of folding increases. To display this phenomenon more clearly, we use values of d that span a larger range (see Fig. 4.6a, f). These results are consistent with the simulation results of Toro et al. [22].

To compare the simulation results across changing values of r_2 and d, ratios of brain radius to cortical thickness were arranged in Table 4.1. When the ratio of brain radius to cortical thickness decreases, the degree of folding also decreases (see Table 4.1).

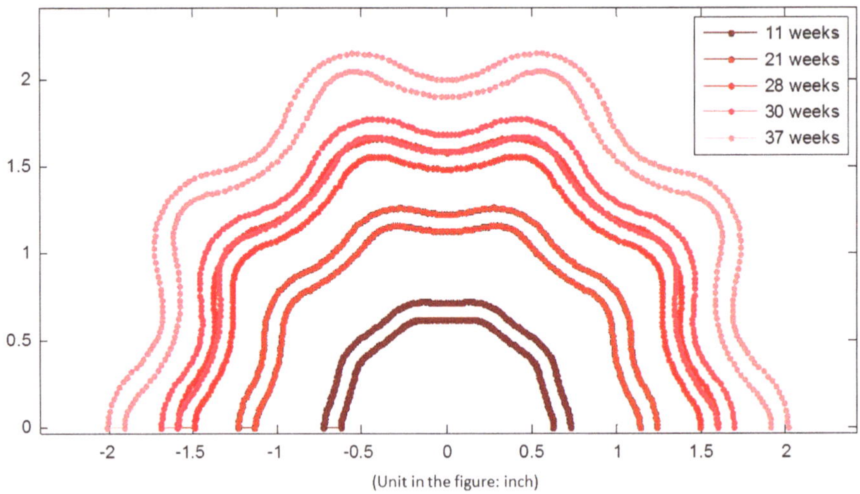

Fig. 4.5 Changing gray matter radius r_2. Values of r_2 represent the gray matter radius taken at various gestational ages (GA) [1, 16]: 11 weeks GA (r_2 = 1.8351 cm), 21 weeks GA (r_2 = 3.1303 cm), 28 weeks GA (r_2 = 4.0464 cm), 30 weeks GA (r_2 = 4.2966 cm), 37 weeks GA (r_2 = 5.1088 cm). All other applied forces and parameters are as in Fig. 4.2b

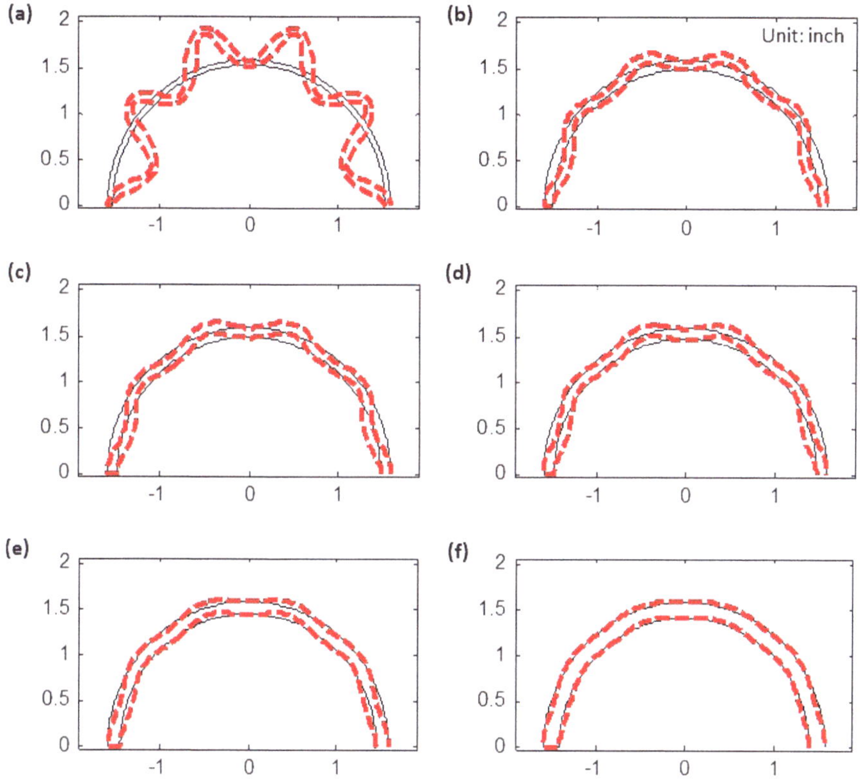

Fig. 4.6 Changing cortical thickness d. **(a)** $d = 1.5$ mm, **(b)** $d = 2.5$ mm (overall average), **(c)** $d = 2.7$ mm (medial cortex), **(d)** $d = 3.0$ mm (inferior cortex), **(e)** $d = 3.5$ mm (lateral cortex), **(f)** $d = 4.5$ mm. All other applied forces and parameters are as in Fig. 4.2b

Table 4.1 Ratio of brain radius to cortical thickness and degree of cortical folding

		Parameters		Ratio	
		r_2 (mm)	d (mm)	r_2/d	Degree of folding
Figure 4.5	11 weeks	18.351	2.5	7.3404	Less folding
	21 weeks	31.303	2.5	12.5212	
	30 weeks	42.966	2.5	17.1864	
	37 weeks	51.088	2.5	20.4352	More folding
Figure 4.6	(a)	40.464	1.5	26.9760	More folding
	(b)	40.464	2.5	16.1856	
	(c)	40.464	2.7	14.9867	
	(d)	40.464	3.0	13.4880	
	(e)	40.464	3.5	11.5611	
	(f)	40.464	4.5	8.9920	Less folding

These simulation results can be interpreted in two different ways. Species with smaller cortices will tend to have a relatively smoother cortex, while those with larger brains will tend to have a highly convoluted cortex when the same tangential forces are applied. This interpretation agrees with that of [25]. An alternative interpretation to these results is that as the brain grows, it is expected that more cortical convolutions will develop.

These results show that it is possible for tangential tension to lead to folding, particularly in larger brains. Interestingly, Xu et al. [28] have used data from developing ferret brains to argue that significant axonal tension does not exist in the cores of developing gyri; rather, considerable tension is present in subcortical white matter regions where it is too deep to drive cortical convolutions. However, our results show that a combination of both cortical thickness, gray matter radius, and tangential tension influence cortical convolutions. Our results confirm that small thickness and small gray matter radius, such as in the ferret, will not exhibit folding due to tangential tension. In contrast, our results indicate that a larger brain radius to cortical thickness ratio, such as in humans, will exhibit folding due to tension (see Fig. 4.5; compare inner convolutions to outer convolutions). Thus, it is possible that tensions due to tangential forces located in the upper white matter regions of the human brain is enough to affect to cortical convolution development in gestation. Therefore, it is not clear if the conclusions regarding folding pattern development in the ferret can be applied to humans. Further experimental investigations are needed.

4.4 Conclusions

In this study, we have proposed a two-dimensional biomechanical model of cortical folding. We assumed the body forces in elasticity theory [4] represent axonal tension. Axonal tension has been suggested as a major mechanism driving cortical convolution [25]. We applied tangential and radial forces at finite element nodes on the boundary between the white matter and gray matter to study the effects of axonal tension. In addition, the effects of folding due to the magnitude of the tension vectors were examined, as were the effects of domain size and thickness.

Our proposed model provides an approach for studying the tension-based hypothesis of cortical folding [25] numerically as well as theoretically. The ability to freely set the direction of the applied tension vectors is an advantage of our model. In addition, our proposed model uses plausible brain parameters. With our model, we are able to show the effects of the magnitude of radial and tangential forces affecting tension. Using realistic biophysical values of brain tissue elasticity parameters, cortical gray matter radius, and cortical thickness, our model can be used to estimate folding depth due to tension. Regarding domain size, in particular cortical thickness and radius of the gray matter, our simulation results are consistent with many previous studies and expectations concerning cortical folding development.

Our model can be improved to be more realistic by modifying it to be three-dimensional and time-dependent on a growing domain. It is natural to expect cortical convolution development in one stage will influence the convolutions in the next developmental time step. A realistic 3D model requires the shape of the brain during development and tissue elasticity parameters (Young's modulus and Poisson's ratio) for a developing brain could also improve the model. In addition, combining other theories with our model will enhance the understanding of the mechanisms underlying cortical folding development. For example, the theory of differential growth of inner and outer layers [17] should contribute to cortical convolution. By combining these theories together, we will be able to investigate cortical convolution development more in detail.

Acknowledgements The authors would like to acknowledge the support of the Institute for Pure and Applied Mathematics (IPAM) at the University of California Los Angeles, the organizers of the program Women in Shape (WiSh): Modeling Boundaries of Objects in 2- and 3-Dimensions that was held at IPAM (in cooperation with Association for Women in Mathematics (AWM)) during July 2013, and the Mathematical Biosciences Institute (MBI) at Ohio State University.

Appendix

Linear Quadrilateral Element

In the non-dimensional coordinate system (ξ, η) (see Fig. 4.7), the four shape functions can be expressed as

$$N_1 = \frac{1}{4}(1-\xi)(1-\eta)\,,\quad N_2 = \frac{1}{4}(1+\xi)(1-\eta)\,,$$

$$N_3 = \frac{1}{4}(1+\xi)(1+\eta)\,,\quad N_4 = \frac{1}{4}(1-\xi)(1+\eta)\,.$$

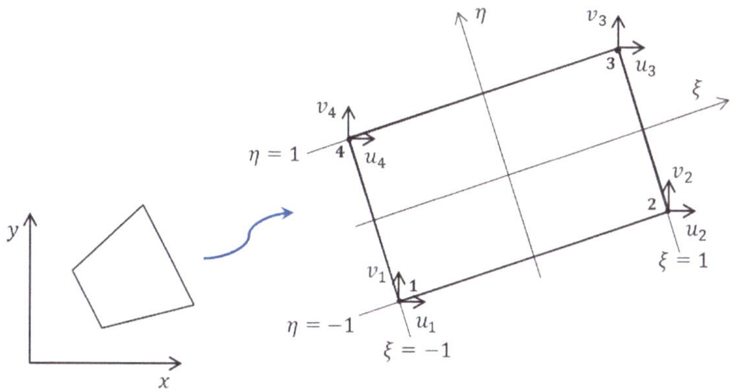

Fig. 4.7 Linear quadrilateral element

At any point inside the element, $\sum_{i=1}^{4} N_i = 1$. The displacement field is expressed as

$$u = \sum_{i=1}^{4} N_i u_i \ , \ \ v = \sum_{i=1}^{4} N_i v_i \ .$$

The shapes of the interpolation functions N_i are twisted planes whose height is 1 at i-th corner of the element and 0 at the other corners. The partial derivatives with respect to the variables are linear functions [4].

References

1. Armstrong, E., Schleicher, A., Omran, H., Curtis, M., Zilles, K.: The ontogeny of human gyrification. Cereb. Cortex **5**(1), 56–63 (1995)
2. Barkovich, A.: Current concepts of polymicrogyria. Neuroradiology **52**, 479–487 (2010)
3. Bayly, P., Okamoto, R., Xu, G., Shi, Y., Taber, L.: A cortical folding model incorporating stress-dependent growth explains gyral wavelengths and stress patterns in the developing brain. Phys. Biol. **10**, 016,005 (2013). doi:10.1088/1478-3975/10/1/016005
4. Bickford, W.: A First Course in the Finite Element Method, 2nd edn. Irwin Publishers, Burr Ridges (1994)
5. Chi, J., Dooling, E., Gilles, F.: Gyral development of the human brain. Ann. Neurol. **1**, 86–93 (1977)
6. Dobyns, W., Curry, C., Hoyme, H., Turlington, L., Ledbetter, D.: Clinical and molecular diagnosis of Miller-Dieker syndrome. Am. J. Hum. Genet. **48**(3), 584–594 (1991)
7. Fischl, B., Dale, A.: Measuring the thickness of the human cerebral cortex from magnetic resonance images. Proc. Natl. Acad. Sci. U.S.A. **97**(20), 11,050–11,055 (2000)
8. Geng, G., Johnston, L., Yan, E., Britto, J., Smith, D., Walker, D., Egan, G.: Biomechanisms for modelling cerebral cortical folding. Med. Image Anal. **13**, 920–930 (2009). doi:10.1016/j.media.2008.12.005
9. Griffin, L.: The intrinsic geometry of the cerebral cortex. J. Theor. Biol. **166**, 261–273 (1994)
10. Hofman, M.: Size and shape of the cerebral cortex in mammals. I. The cortical surface. Brain Behav. Evol. **27**(1), 28–40 (1985). doi:10.1159/000118718
11. Im, K., Lee, J.M., Lyttelton, O., Kim, S., Evans, A., Kim, S.: Brain size and cortical structure in the adult human brain. Cereb. Cortex **18**, 2181–2191 (2008). doi:10.1093/cercor/bhm244
12. Kriegstein, A., Noctor, S., Martinez-Cerdeno, V.: Patterns of neural stem and progenitor cell division may underlie evolutionary cortical expansion. Nat. Rev. Neurosci. **7**, 883–890 (2006). doi:10.1038/nrn2008
13. Kwon, Y., Bang, H.: The Finite Element Method Using MATLAB, 2nd edn. CRC, Boca Raton, Flordia (2000)
14. Lemaitre, H., Goldman, A., Sambataro, F., Verchinski, B., Meyer-Lindenberg, A., Weinberger, D., Mattay, V.: Normal age-related brain morphometric changes: Nonuniformity across cortical thickness, surface area and grey matter volume? Neurobiol. Aging **33**(3), 617.e1–617.e9 (2012)
15. Nie, J., Guo, L., Li, G., Faraco, C., Miller, L., Liu, T.: A computational model of cerebral cortex folding. Cereb. Cortex **15**(12), 1900–1913 (2005). doi:10.1016/j.jtbi.2010.02.002
16. Raghavan, R., Lawton, W., Ranjan, S., Viswanathan, R.: A continuum mechanics-based model for cortical growth. J. Theor. Biol. **187**(2), 285–296 (1997)
17. Richman, D., Stewart, R., Hutchinson, J., Caviness, V.: Mechanical model of brain convolutional development. Science **189**, 18–21 (1975)

18. Soza, G., Grosso, R., Nimsky, C., Hastreiter, P., Fahlbusch, R., Greiner, G.: Determination of the elasticity parameters of brain tissue with combined simulation and registration. Int. J. Med. Robot. **1**(3), 87–95 (2005)
19. Striegel, D., Hurdal, M.: Chemically based mathematical model for development of cerebral cortical folding patterns. PLoS Comput. Biol. **5**(9), e1000524 (2009)
20. Taylor, Z., Miller, K.: Reassessment of brain elasticity for analysis of biomechanisms of hydrocephalus. J. Biomech. **37**(8), 1263–1269 (2004)
21. Toole, G., Hurdal, M.: Turing models of cortical folding on exponentially and logistically growing domains. Comput. Math. Appl. **66**(9), 1627–1642 (2013)
22. Toro, R., Burnod, Y.: A morphogenetic model for the development of cortical convolutions. Cereb. Cortex **15**(12), 1900–1913 (2005)
23. Tyler, W.: The mechanobiology of brain function. Nat. Rev. Neurosci. **13**, 867–878 (2012)
24. University of Wisconsin and Michigan State Comparative Mammalian Brain Collections: Comparative mammalian brain collections. Website: http://www.brainmuseum.org/. Last accessed: 10 Jan 2014
25. Van Essen, D.: A tension-based theory of morphogenesis and compact wiring in the central nervous system. Nature **385**(6614), 313–318 (1997)
26. Wisniewski, K.: Down syndrome children often have brain with maturation delay, retardation of growth, and cortical dysgenesis. Am. J. Med. Genet. Suppl. **7**, 274–281 (1990)
27. Wittek, A., Hawkins, T., Miller, K.: On the unimportance of constitutive models in computing brain deformation for image-guided surgery. Biomech. Model. Mechanobiol. **8**(1), 77–84 (2009)
28. Xu, G., Knutsen, A., Dikranian, K., Kroenke, C., Bayly, P., Taber, L.: Axons pull on the brain, but tension does not drive cortical folding. J. Biomech. Eng. **132**(7), 071,013 (2010). doi:10.1115/1.4001683

Chapter 5
Quantification and Visualization of Variation in Anatomical Trees

Nina Amenta, Manasi Datar, Asger Dirksen, Marleen de Bruijne, Aasa Feragen, Xiaoyin Ge, Jesper Holst Pedersen, Marylesa Howard, Megan Owen, Jens Petersen, Jie Shi, and Qiuping Xu

N. Amenta
University of California at Davis, Davis, CA, USA

M. Datar
Scientific Computing and Imaging Institute, University of Utah, Salt Lake City, UT, USA

A. Dirksen
Lungemedicinsk Afdeling, Gentofte Hospital, Hellerup, Denmark

M. de Bruijne
Department of Computer Science, University of Copenhagen, Copenhagen, Denmark
Erasmus MC, Rotterdam, The Netherlands

A. Feragen (✉) • J. Petersen
Department of Computer Science, University of Copenhagen, Copenhagen, Denmark
e-mail: aasa@diku.dk

X. Ge
Department of Computer Science and Engineering, Ohio State University, Columbus, OH, USA

J.H. Pedersen
Department of Cardiothoracic Surgery, Rigshospitalet, University of Copenhagen, Copenhagen, Denmark

M. Howard
National Security Technologies, LLC (A Department of Energy Contractor), Las Vegas, NV, USA

M. Owen (✉)
Department of Mathematics and Computer Science, Lehman College, City University of New York, Bronx, NY, USA
e-mail: megan.owen@lehman.cuny.edu

J. Shi
School of Computing, Informatics, and Decision Systems Engineering, Arizona State University, Tucson, AZ, USA

Q. Xu
Department of Mathematics, Florida State University, Tallahassee, FL, USA

K. Leonard, S. Tari (eds.), *Research in Shape Modeling*, Association for Women in Mathematics Series 1, DOI 10.1007/978-3-319-16348-2_5

Abstract This paper presents two approaches to quantifying and visualizing variation in datasets of trees. The first approach localizes subtrees in which significant population differences are found through hypothesis testing and sparse classifiers on subtree features. The second approach visualizes the global metric structure of datasets through low-distortion embedding into hyperbolic planes in the style of multidimensional scaling. A case study is made on a dataset of airway trees in relation to Chronic Obstructive Pulmonary Disease.

5.1 Introduction

Tree-structured data appears in many medical imaging applications, e.g., airway trees [14], blood vessel trees [18], dendrites [32] and galactograms [26]. Typically, these anatomical trees vary both in tree topology and associated branch features such as branch length or shape, and as a result there is no straight-forward way to analyze the trees using standard Euclidean statistics. One way to integrate both tree topology and branch features in a single parametric framework is by modeling trees as residing in a non-linear, non-smooth *tree-space* [7, 14]. The non-linear, non-smooth nature of tree-space creates several problems for data analysis. First, statistics have to be redefined, as the standard statistical procedures such as finding an average or a principal component, or performing classification, do not translate directly to the tree-space setting. Second, even if we define classification algorithms in tree-space, we do not know which parts of the anatomical tree are responsible for causing class differences, for example, because each tree-space point represents an entire tree structure. Third, due to the lack of statistical tools such as principal component analysis, it is hard to visualize how distributions of trees vary in tree-space. While recent work has resulted in basic statistical tools [5, 14, 27, 29], the two latter problems are still unsolved. In this paper we investigate two approaches to these two problems: First, we study the influence of local subtrees on the results of hypothesis testing and classification, and the identification of subtrees which are responsible for significant differences between two populations of trees. Second, we use *hyperbolic low-distortion embedding* to visualize the global metric structure of data living in tree-space. As a case study, we demonstrate the use of these techniques on a population of airway trees from a lung cancer screening study.

This paper presents results from the 1-week collaboration workshop *Women in Shape: Modeling Boundaries of Objects in* 2- *and* 3-*Dimensions* held at the Institute of Pure and Applied Mathematics at UCLA, July 15–19 2013. At this workshop, most of the authors of this paper spent a week working together on two projects related to quantifying and visualizing variance in populations of trees, which are described in Sects. 5.3 and 5.4, respectively.

5.1.1 Tree-Space

A tree-space is any geometric space in which points represent trees. The tree-space used in this paper, described in [14, 15], is a generalization of the phylogenetic tree-space proposed by Billera et al. [7]. This tree-space, denoted $\mathcal{T}_n$, contains all rooted trees with n labeled leaves with vertices of degree at least 3, where the n leaf labels are given by a fixed set of cardinality n. In this paper, the root of a tree is not considered to be among the leaves. Furthermore, for any tree in this tree-space, each edge has a k-dimensional vector associated with it. An example of such an edge vector is a non-negative real number representing the edge length (i.e. $k = 1$); a second example is the vector of l 3-dimensional landmark points sampled along a branch centerline, giving $k = 3 \cdot l$. For each edge, the landmark points are translated so that the edge starts at the origin. We refer to the latter edge vector as the *shape* of the edge. The trees in such a tree-space can, for instance, be used to model airways in the lung. In this space, we will use edge shape with $l = 5$ to describe edges unless otherwise stated.

We now give a description of our tree-space, $\mathcal{T}_n$, for a fixed n, which is illustrated for $n = 4$ in Fig. 5.1. All the trees in $\mathcal{T}_n$ with the same *tree topology*, or branching order, form a lower-dimension Euclidean subspace in the tree-space. The dimension of this subspace is mk, where m is the number of edges in the tree topology and k is the dimension of the vector associated with each tree edge. Each tree edge in the tree topology is put into correspondence with k of the subspace's dimensions, and a particular tree with that topology can be written as a km-dimensional vector in that subspace, with the coordinates being the consecutive k-dimensional edge vectors. That is, if a tree has edges $e_1, e_2, \dots, e_m$, with corresponding edge vectors $\boldsymbol{\ell}_1 = (\ell_1^1, \ell_2^1, \dots, \ell_5^1), \dots \boldsymbol{\ell}_\mathbf{m} = (\ell_m^1, \ell_m^2, \dots, \ell_m^5)$, then that tree corresponds to the

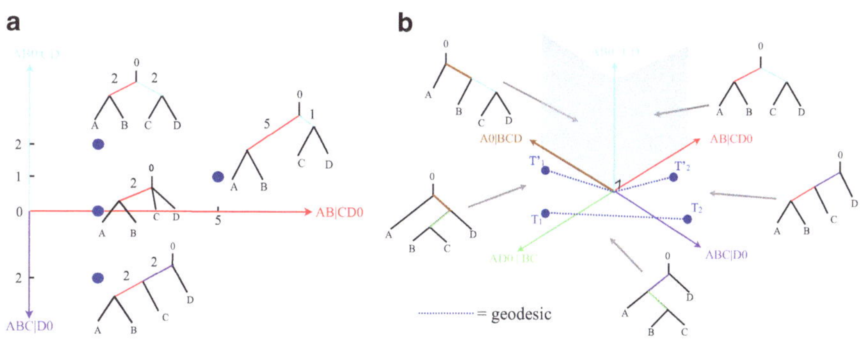

Fig. 5.1 (**a**) Two adjacent quadrants in the tree space, $\mathcal{T}_4$, of trees with 4 leaves and edge vectors of 1 dimension. Here, the edge vector, or length, is restricted to being non-negative and the pendant edges are ignored, so the Euclidean subspaces are represented as quadrants. A quadrant contains all trees with a given topology, and each tree with that topology is represented by the coordinates corresponding to its internal edge lengths. (**b**) Representation of 5 of the 15 quadrants in the tree space $\mathcal{T}_4$, with edge vectors of 1 dimension

point $(\boldsymbol{\ell}_1, \ldots, \boldsymbol{\ell}_m)$. All trees within the subspace must map their k-dimensional edge vectors to the km-dimensional vector in the same order, but what this order is does not matter.

The Euclidean subspaces for each tree topology are glued together in the following way. Consider a tree containing the edges $e_1, \ldots, e_m$, where each edge can be identified by the unique partition of the leaves it makes when it is removed from the tree (i.e. removing the edge forms a forest of two trees, each of whose leaves, including the root, forms one half of the partition). Let exactly one of the edges, say e_1, have an all 0 edge vector. Then this tree lies on the boundary of the Euclidean subspace corresponding to its tree topology, and furthermore, it actually lies in a lower dimensional Euclidean subspace E corresponding to trees with only the edges $e_2, \ldots, e_m$ in their topologies. This lower dimensional subspace E is also on the boundary of two other Euclidean subspaces, and we identify all such common subspaces in all the Euclidean subspaces corresponding to tree topologies to form $\mathcal{T}_n$. See [15] and [7] for a more detailed description of the tree space.

The metric on $\mathcal{T}_n$ is induced by the Euclidean metric on each of its constituent subspaces. Specifically, the distance between two trees with the same topology is the Euclidean distance between the two points representing those trees in the subspace for that tree topology. The distance between two trees with different topologies is the length of the shortest path joining their points in tree space. Such a path will consist of a sequence of line segments, each contained in exactly one of the subspaces, and thus the path length is just the sum of the Euclidean lengths of each segment.

Most importantly for this paper, tree space is a non-positively curved metric space [7], which implies that there is a unique shortest path within the space between any two trees, called the *geodesic*. The *geodesic distance* between two trees is the length of the geodesic between them, and it can be computed in polynomial time [30]. Certain statistics, such as means and first principal components, can also be computed on trees in this space [5, 14, 27, 29].

While sections of tree-space are identical to higher dimensional Euclidean spaces, tree-space itself is not a manifold. In particular, it has several singularities, which have infinite negative curvature. One such singularity is at the origin, which corresponds to the tree which has all 0 edge lengths or vectors. A simpler model of this singularity in $\mathcal{T}_4$ is a *corner*: five Euclidean quadrants, glued together around an single origin, see Fig. 5.1b. Singular points in tree-space also occur where the higher dimensional Euclidean subspaces join together to form a space that locally resembles an *open book*. An open book is a set of Euclidean half planes, or *sheets*, which are identified along their boundary hyperplanes, which form the *spine*, see Fig. 5.2.

5.1.2 The Fréchet Mean in Tree-Space

In Euclidean space, there are a number of equivalent definitions of the mean. Some of these definitions cannot be carried over to tree-space, while those that

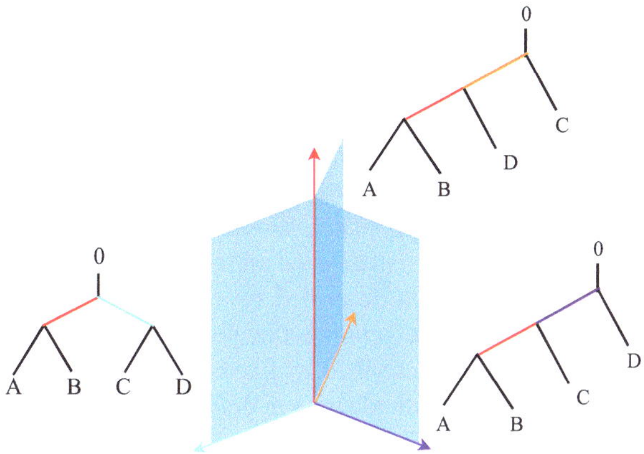

Fig. 5.2 An occurrence of a half open book with 3 2-dimensional sheets in tree-space

can be carried over are no longer equivalent. One of the definitions of mean in Euclidean space is the Fréchet mean, or barycenter, which minimizes the sum of square distances to the input set. If $\{T_1, T_2, \ldots, T_r\}$ are a set of input trees, then their Fréchet mean in tree-space is the tree t which minimizes $\sum_{i=1}^{r} d(t, T_i)^2$, where d is the geodesic distance. The Fréchet mean was introduced for tree-space independently by [5] and [27], both of whom also gave an algorithm to approximate it based on a Law of Large Numbers holding in non-positively curved spaces [35].

5.2 Related Work

This paper studies two problems related to understanding variance in datasets of trees: (i) detection of local subtree differences, and (ii) visualization of global population-level geometry.

5.2.1 Local Significant Differences

Many data types represent entities which can be decomposed into parts or regions. Examples are graph-structured data [13, 23], anatomical data which can be segmented into different organs [17, 20] or even single anatomical organs where additional spatial information is relevant; for instance, in the framework of shape analysis [9, 10] where local analysis is made on correspondence points on biomedical shape surfaces. A typical problem when studying such data is *interpretability*: A classifier will often only predict a certain diagnosis or class, but in order to understand the cause of the result (and, e.g. in diagnostic settings, react on it) one also desires to know which parts of the collection caused a certain classification outcome.

While a large body of work has been done on classifying structured data, less is known about how to identify which parts of a structure are relevant for the classification problem. Most such work has been done in settings where there is a correspondence between the parts constituting the data object: In analysis of brain connectivity [19, 23], one usually has a matching between the nodes in the dataset, while in voxel-based morphometry [3] or shape analysis [10], registration is used to match different images to a template. A popular approach to such problems is *structured sparsity* [4, 22, 23], which detects discriminative substructures in data described by fixed-length Euclidean vectors with a known underlying structure relating the vector coordinates. However, anatomical trees usually cannot be described by fixed-length vectors without discarding parts of the tree. Thus, these methods are not directly applicable.

5.2.2 Low-Distortion Embeddings

The standard technique for visualising population structure in high-dimensional or non-Euclidean datasets is to extract the pairwise distances between data points, and then use *multidimensional scaling* (MDS), which attempts to embed the points into a lower dimensional Euclidean space such that the given distances between the points are preserved. This is expressed mathematically as minimizing the sum of the differences between original and embedded pairwise distances [8]. In a sequence of work [2, 21, 28] Amenta, St. John et al. investigate visualization of sets of phylogenetic, or evolutionary, trees using multidimensional scaling. In this work, inter-tree distances are given by the Robinson-Foulds distance [34], which only measures topological differences in the trees. More recently, Wilgenbusch et al. [41] compare several non-linear versions of MDS on phylogenetic trees, and find that a metric that places less weight on large distances gives more meaningful visualizations. Chakerian and Holmes [11] use MDS with the geodesic distance between trees [7]. A different approach is that of Sundberg et al. [36], who visualize phylogenetic trees by projecting them onto a hypersphere; this approach does not consider branch lengths, only tree topology.

All of these methods approach visualization through embedding into a Euclidean space in a low-distortion way. However, embedding spaces need not be restricted to only Euclidean spaces. For instance, low-distortion embedding of a general metric into a tree has been considered for various measures of distortion [1, 6]. Low-distortion embedding of general metrics into hyperbolic spaces has also been considered by Walter et al. [39, 40] and Cvetkovski and Crovella [12]. In this paper, we use hyperbolic MDS for more truthful visualizations of tree variation.

5.3 Quantification and Visualization of Local Tree-Shape Differences

While previous work [14] developed methods for finding significant differences between populations of trees, this work did not address the question of where these changes came from. In this section we investigate different methods for detecting *where* in a tree significant differences appear. In Sect. 5.3.1 we perform hypothesis testing on nested subtrees in order to detect how significant changes take place in particular subtrees. In Sect. 5.3.2 we develop a structured sparsity framework which takes advantage of the tree-space geometry in order to handle the fact that subtrees have variable topological structure. In both of these sections, we obtain results on which subtrees induce significant differences. A disadvantage of the methods developed in Sects. 5.3.1 and 5.3.2 is that they do not take correlation between different subtrees into account. In Sect. 5.3.3 we therefore develop a method that allows us to study how subtree differences correlate with each other.

Case study. We apply the developed methods to a case study of airway trees from subjects with and without Chronic Obstructive Pulmonary Disease (COPD). The 600 airway trees are from randomly selected subjects from the Danish Lung Cancer Screening Trial [31], of which 300 were diagnosed with COPD at scan time and 300 were symptom free. The hypothesis testing and classification experiments performed in this chapter all have the common goal of separating the class of COPD patients from the class of symptom free subjects.

The airway trees were extracted from low-dose (120 kV and 40 mAs) pulmonary CT scans. To extract the tree, the airway lumen surface was extracted from the images using the locally optimal path approach of [24] and then refined using the optimal surface approach of [33]. Afterwards centerlines were computed by front propagation within the refined lumen surface as described in [25]. The resulting centerlines were disconnected in bifurcation regions and so Dijkstra's algorithm was used to connect them along shortest paths within an inverted distance transform of the refined lumen surface. These centerlines were then represented by 6 equidistantly sampled landmark points. The airway trees were normalized by patient height as an affine scaling parameter.

Airway trees are somewhat regular in the sense that some of the branches have anatomical names and can be found in most human lungs. The subtrees rooted at these branches feed different subdivisions of the lung at different hierarchical levels, as schematically illustrated in Fig. 5.3. The Trachea is the root branch that feeds both lungs. The left and right main bronchi (LMB, RMB) feed the left and right lungs. The left upper lobe and lower lobe branches (LUL and LLB) feed the left upper and lower lobes. The lower lobe splits into the branches L7–L10. The left upper lobe branches into two subsections; the first feeds the three segments L1, L2 and L3, and the branch feeding all of these is called L1+2+3. The second subsection feeds the segments L4 and L5, and their parent is called L4+5. The right lung us subdivided into the upper lobe, fed by the right upper lobe branch (RUL), and the middle- and

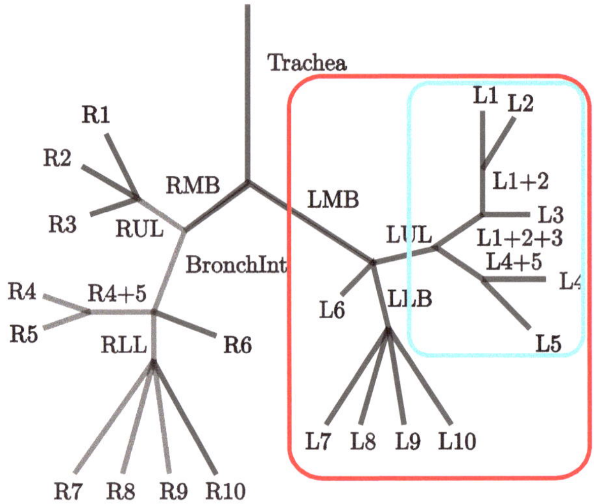

Fig. 5.3 Airway tree (*black*) and sub-trees (LMB, *red*) and (LUL, *blue*)

lower lobes, both fed by the bronchus intermedius (BronchInt). The middle lobe consists of the segments R4–R5, fed by the parent R4+5, and the lower lobe is fed by the right lower lobe branch (RLL), and splits into the segments R7–R10.

Due to variation in airway tree topology, these branches are not always all present. In our dataset, however, which has been automatically labeled using the algorithm presented in [15], the following branches are consistently present:

$$\text{Trachea, LMB, RMB, LUL, RUL, L1+2+3, LLB, BronchInt, and RLL.} \quad (5.1)$$

5.3.1 Permutation Tests for Subtree Statistics

In this section we perform subtree hypothesis testing using the tree-shape permutation tests for equality of means and variances developed in [14] on the nested subtrees defined by the subtree root branches (5.1). These tests are standard permutation tests which, for samples G_1 and G_2 drawn from two different classes of trees (in our experiments: healthy subjects and COPD patients), use test statistics used for means and variances between classes defined as

$$t_m = d(\mu(G_1), \mu(G_2)) \text{ and } t_v = |v(G_1) - v(G_2)|,$$

respectively, where $\mu(G_i)$ is the Fréchet mean of the trees in the ith class as defined on p. 60, and $v(G_i)$ denotes the variance of the ith class $\frac{1}{(N-1)} \sum_{t \in G_i} d^2(t, \mu(G_i))$.

Under the null hypothesis, namely that there is no difference between the two classes, the samples G_1 and G_2 are drawn from the same distribution on $\mathcal{T}_n$, and randomly permuting the elements of G_1 and G_2 should not affect the value of the test statistic t_*.

Form the two-class data set $G = G_1 \cup G_2$ and consider partitions of G into subsets of size $N_1 = |G_1|$ and $N_2 = |G_2|$. Due to the size of G we cannot check all possible permutations, but instead compute the test statistics $(t_*)_m$ for means and variances for the new subsets, $m = 1, \ldots, M$, for M random partitions of G into sets of size N_1 and N_2. Comparing the $(t_*)_m$ to the original statistic value t_* for the samples G_1 and G_2, we obtain a p-value approximating the probability of observing t_* under the null hypothesis:

$$p = \frac{1 + \sum_{(t_*)_m \geq t_*, m \in \{1,\ldots,M\}} 1}{M + 1},$$

where the additional 1 is added to avoid $p = 0$.

Permutation tests for the two statistics were performed on each subtree with $M = 1{,}000$ permutations. The results are summarized in Table 5.1, and show significant differences in several subtrees. These subtrees are identified by their root branches, which are illustrated schematically in Fig. 5.3. In comparison, the same hypothesis test was made on the individual branches (5.1) along with the segment branches R1–R10, L1–L10, with results shown in Table 5.2. The tests on individually identified branches show relatively fewer significant differences between the two populations, emphasizing a need for considering the airway subtrees as entities rather than collections of independent branches.

This suggests that the permutation test described here can be applied to study local group differences between subtrees in a hierarchical manner.

Table 5.1 Case study: Group comparison showing results of permutation tests on subtrees of the full airway trees rooted at the branches listed in (5.1). The permutation test compares the populations of airway trees from COPD patients and symptom free subjects

Label	P-value mean	P-value variance
Full	**0.0010**	**0.0060**
RMB	**0.0020**	0.0939
RUL	0.2298	0.1668
BrInt	**0.0050**	0.1249
RLL	**0.0300**	0.0959
LMB	0.0859	**0.0210**
LUL	**0.0320**	**0.0390**
L123	**0.0260**	**0.0410**
LLB	0.5524	0.1588

Table 5.2 Case study benchmark: Permutation tests for shape differences in individual branches between populations of airway trees from healthy individuals and COPD patients. P-values below a threshold of 0.05 are shown in bold

Label	P-value mean	P-value variance	Label	P-value mean	P-value variance
RMB	0.163	0.621	LMB	**0.020**	0.786
RUL	0.134	0.416	LUL	0.297	0.118
R1	0.410	0.363	L1	0.163	0.391
R2	0.116	0.255	L2	0.324	**0.017**
R3	0.329	0.854	L3	0.968	0.800
BronchInt	**0.001**	0.764	L45	0.078	0.312
R4	0.134	0.190	L4	0.372	0.570
R5	**0.027**	0.175	L5	**0.023**	**0.050**
R6	0.992	0.135	L6	0.260	0.833
RLL	**0.001**	0.865	LLB	0.177	0.112
R7	0.058	0.325	L7	0.496	0.611
R8	**0.014**	0.207	L8	0.466	0.900
R9	**0.037**	0.127	L9	0.146	**0.026**
R10	0.308	0.652	L10	0.855	0.162
L123	0.393	0.361			

5.3.2 *Subtree Classification*

In the previous section we saw how hypothesis testing on subtrees allowed us to learn about which subtrees differed significantly between two populations of trees. While significant differences are interesting in their own right, we are often particularly interested in finding *predictive* differences. In particular, we want to find subtrees such that restricting prediction to these subtrees results in good predictive performance, giving interpretable classifiers in the sense that we can detect *which* tree changes are predictive.

5.3.2.1 Classification on Known Branches

A straight-forward approach to tree classification and identification of discriminative substructures of trees is to use standard classification methods on vectors whose coordinates correspond to a fixed set of identified branches. In our case, these branches will be identified by their anatomical names, which create a natural matching between the branches of different trees. Classification of such vectors can return information about which branches are more discriminative, because classifiers such as the support vector machine (SVM), include coordinate weights that intuitively correspond to the relevance of the individual coordinate feature for the classification problem. This method is simple, but has the disadvantage that it can only use branches that are present in every single tree in the dataset. This method will form a baseline to which our proposed methods are compared.

Table 5.3 Mean ± standard deviation of COPD classification accuracy using branch length (left) and branch shape (right)

Method	Accuracy length	Accuracy shape
LDA	0.56 ± 0.06	0.52 ± 0.06
QDA	0.55 ± 0.05	N/A
Mahalanobis	0.54 ± 0.05	N/A
kNN	0.53 ± 0.06	0.53 ± 0.06
SVM	0.56 ± 0.06	0.56 ± 0.06

Table 5.4 The mean and standard deviations of the SVM weight vectors on the COPD/healthy classification. The largest weight vectors as well as those falling within one standard deviation are highlighted

Branch	SVM weight	Branch	SVM weight	Branch	SVM weight	Branch	SVM weight
RMB	2.0 ± 1.45	R5	2.8 ± 1.6	LMB	-2.6 ± 1.6	L4	-1.4 ± 1.2
RUL	$\mathbf{4.3 \pm 1.6}$	R6	-1.7 ± 1.6	LUL	1.7 ± 2.1	L5	2.3 ± 1.6
R1	-0.7 ± 1.2	L6	1.5 ± 1.6	L123	-0.8 ± 1.4	RLL	2.6 ± 1.6
R2	$\mathbf{4.9 \pm 1.2}$	R7	1.4 ± 1.5	L1	-1.6 ± 1.2	LLB	$\mathbf{-3.6 \pm 1.5}$
R3	$\mathbf{-3.6 \pm 1.4}$	R8	3.3 ± 2.2	L2	$\mathbf{-3.2 \pm 1.3}$	L7	-2.8 ± 1.3
BrInt	$\mathbf{-5.0 \pm 1.6}$	R9	$\mathbf{7.3 \pm 1.5}$	L3	1.4 ± 1.3	L8	2.2 ± 1.5
R4	-0.4 ± 1.6	R10	2.9 ± 1.6	L45	$\mathbf{3.6 \pm 1.8}$	L9	$\mathbf{3.2 \pm 1.5}$
						L10	$\mathbf{5.2 \pm 1.5}$

Since we could only use branches that were present in every single dataset tree, we used the list of branches (5.1) along with the leaves {R1–R10, L1–L10}, which are guaranteed to be present. We performed classification with linear discriminant analysis (LDA), quadratic discriminant analysis (QDA), Mahalanobis distance, k-nearest neighbor (kNN) using $k = 5$, and support vector machine (SVM) using 10 repetitions of 10-cross validation. The corresponding classification accuracies are reported in Table 5.3. Note that the QDA and Mahalanobis distance are missing for the shape branch features; this is because these both require a positive definite covariance matrix, for which the data set was too small for the higher-dimensional shape vectors.

While the mean classification accuracy is above chance for all classifiers, none of them are significantly above chance. A common heuristic to find features which are important in classification is to study the magnitudes of the coordinates of the weight vector produced by the SVM algorithm, as shown in Table 5.4. Note that in addition to the classification accuracy being very low, the weights of high magnitude are scattered around the airway tree, not adding much in terms of interpretation.

The poor performance of classifiers on the set of all branches could be explained by the fact that many branches are highly correlated. In such cases, the weights might not carry much information. The poor classification accuracy may be explained by the dimensionality of the data. This motivates our search for a more predictive and interpretable classification algorithm by including subtree information in the classifier.

5.3.2.2 Structured Sparse Feature Selection Through Regularized Logistic Regression on Subtree Similarity

Under a hypothesis that significant differences are found in local subtrees, we incorporate local subtree structure into classification through a sparse classifier taking subtree similarity as input. Logistic regression measures the relationship between a categorical dependent variable (class label) and one or more independent variables by using conditional probabilities as predicted values of the dependent variable. The L_1 regularized logistic regression, or the so-called sparse logistic regression [38], regularizes the classifier by forcing the weight vector of the classifier to have a small number of nonzero values. This results in implicit feature selection and robustness to noise, as well as interpretability through the selected subtree features. In addition to its solid theoretical foundation, this model is computationally efficient [16].

Consider a set of n training examples $T = \{(x_1, y_1), (x_2, y_2), \cdots, (x_n, y_n)\}$ from which a tree classifier $y = f(x)$ will be learned. The ith tree is represented by a D-dimensional feature vector $x_i = [x_{i1}, x_{i2}, \cdots, x_{id}]^T$ where $x_{ij} = d(S^i_j, \mu_j)$, where S^i_j is the jth subtree of the ith tree, μ_j is the Fréchet mean tree of all jth subtrees in the training set, and d denotes geodesic distance between trees. The D subtrees are rooted at the branches listed in (5.1). The values $y_i \in \{0, 1\}$ indicate the class labels of the two groups, modeling the conditional probability distribution of the class label y given a feature vector x as:

$$p(y = 1|x, \beta) = \frac{1}{1 + \exp(-\beta^T x)},$$

where $\beta \in \mathrm{R}^D$ are the parameters of logistic model. The estimation of the parameters β is done by likelihood maximization, equivalent to minimizing the negative log-likelihood

$$\hat{\beta} = \arg\min \sum_{i=1}^{n} - \log p(y_i|x_i, \beta).$$

Applying a sparse regularizer we obtain feature selection, interpretability and reduced overfitting. This is done by adding a so-called *lasso* regularization term:

$$\hat{\beta} = \arg\min \sum_{i=1}^{n} - \log p(y_i|x_i, \beta) + \lambda ||\beta||_1,$$

where λ is a parameter controlling the sparsity of β, in the sense that fewer nonzero coefficients of β remain as λ increases. The optimal λ is chosen to optimize classification accuracy by 5-fold cross validation.

The nested subtrees used will be correlated by definition. One way of handling this is by adding an l_2 norm regularization term as well, known as *elastic net regularization*. This leads to an objective function

Table 5.5 Results of the structured logistic classifier with lasso and elastic net regularization. Classification results are averaged over 10 randomized folds, and significant features are those where the distances to both class means were kept as features in the classifier in all folds

Value of α	Classification result	Significant features
1 (lasso)	$65 \pm 2.7\,\%$	Full
0.75	$64.3 \pm 2.1\,\%$	RMB, full
0.5	$62.5 \pm 2.4\,\%$	RMB, BronchInt, full
0.25	$62.5 \pm 2.3\,\%$	RMB, LMB, LLB, BronchInt, RLL, full

$$\hat{\beta} = \arg\min \sum_{i=1}^{n} -\log p(y_i|x_i,\beta) + \lambda\left(\alpha||\beta||_1 + \frac{1-\alpha}{2}||\beta||_2^2\right).$$

The results of the sparse classifiers for different values of α are shown in Table 5.5. Note that the classification performance is significantly better than that of the standard classifiers on identified branches seen in Table 5.3. Moreover, note the discriminative subtrees selected by the classifier. The fact that the lasso regularizer results in only the full tree being selected is most likely a result of the correlation between subtrees. As an l_2 regularizer is also added, we obtain a tradeoff between sparsity and including correlated subtrees.

5.3.3 *Subtree Variance Correlation Testing*

As the localized methods use features extracted from nested subtrees, we expect a high degree of correlation between overlapping subtrees. Most of the previously described methods do not take such correlations into account, and this may, in particular, be a problem for the interpretability through selected features. Moreover, it is interesting to know whether variation in non-overlapping subtrees is correlated. In this section we provide a method for testing the correlation between variance in the subtrees. We use the notation from Sect. 5.3.2.2.

To compare the variance between subtrees in the same airway tree, we use the distance from the j-th subtree S_j^i in tree i to the population mean μ_j of all j-th subtrees in some class as a measure of the amount of variation in that subtree S_j^i. We compute the correlation between these distances, represented by the random variable X and Y, for each pair of subtrees (j,k), and measure whether deviation from the mean subtree μ_j is correlated with deviation from the mean subtree μ_k. To measure the correlation, we use Pearson's sample correlation coefficient,

$$r_{xy} = \frac{\sum_{i=1}^{n}(x_i - \bar{x})(y_i - \bar{y})}{(n-1)s_x s_y},$$

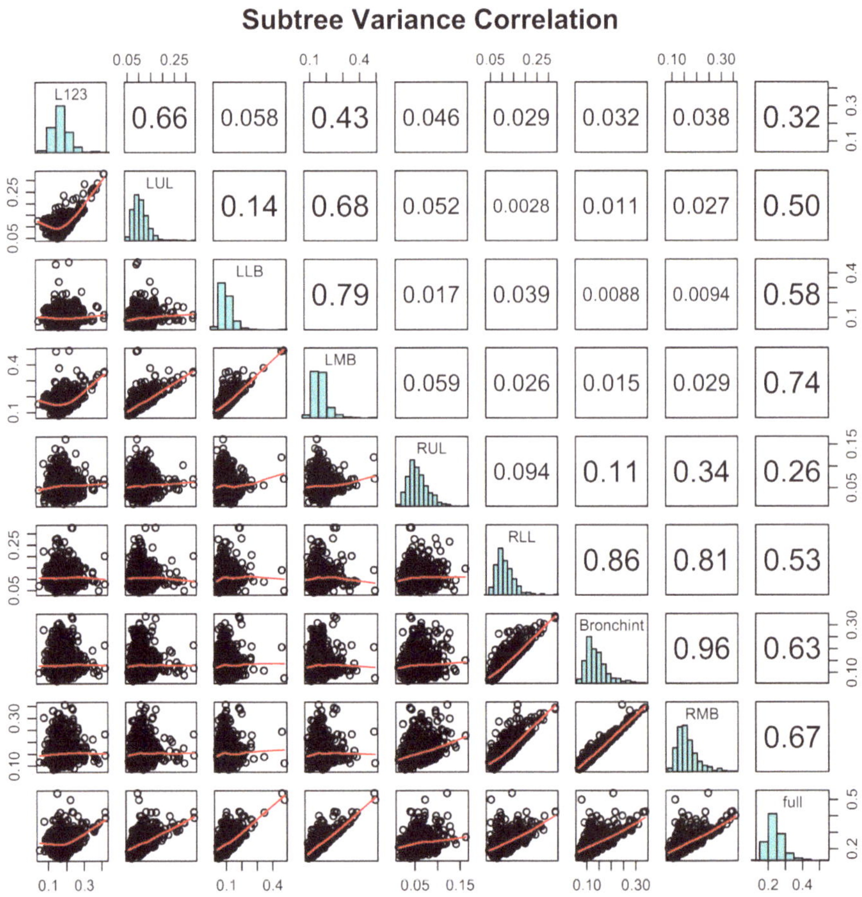

Fig. 5.4 The plots in the lower triangle compare the distances between subtrees in the same tree to their corresponding mean subtree of healthy patients. The plots along the diagonal are the histograms of these distances when the subtree is fixed. The upper diagonal gives the correlation of the distances plotted in the corresponding plot in the lower triangle

where $\bar{x}$ and $\bar{y}$ are the same means of the two distance variables X and Y, and s_x and s_y are the sample standard deviations of the X and Y. This is equivalent to the sample covariance divided by the sample standard deviations.

The results of applying this test to the airway data set is shown in Fig. 5.4. Many of the subtree pairs exhibiting correlation in their variance are nested, as expected. For example, variation is very correlated between the three subtrees RMB, Bronchint, and RLL, where RLL is a subtree of Bronchint, which is itself a subtree of RMB. Similarly, there is a high correlation in the variation between the nested pairs LMB and LUL, and LMB and LLB. However, not all nested subtrees are highly correlated. In particular, RUL is also a subtree of RMB, but variation in it is not very correlated with that in RMB – in fact, variance in RUL is not strongly correlated with any other subtree. The trees are not separated by class as there was no significant difference in the behavior of the two classes.

5.4 Visualization of NPC Information Spaces via Low-Distortion Embedding into the Hyperbolic Disc

High-dimensional data is often embedded into lower dimensional spaces in order to improve the efficiency of computations, or, with a two- or three-dimensional embedding space, for visualization. Besides being high-dimensional and stratified, the tree space $\mathcal{T}_n$ has negative curvature. While it is exactly this property that gives it unique geodesics, it also means that the number of trees within a given neighborhood can grow exponentially with the radius of the neighborhood. Our hypothesis was that embedding point sets in $\mathcal{T}_n$ into lower-dimensional hyperbolic space, which also has negative curvature, would allow embeddings with lower distortion and/or lower total error. We explore the use of two different visualization techniques for general metric distance matrices, Multi-Dimensional Scaling and Isomap.

Multidimensional Scaling (MDS) is a classical approach that maps the original metric dataset to a target Euclidean space, usually of low dimension. It transforms the input metric distance matrix into a set of coordinate positions for the data points – in our case, each tree is a data point – such that the Euclidean distances between the coordinates approximate the input distances as well as possible. Using the new embedded coordinates, one can visualize dataset structure through the embedded dataset, where inter-point distances have been preserved as well as possible. Different definitions for what it means to preserve the distances "as well as possible" produce different computational problems. Principle Components Analysis (PCA) can be seen as a version of MDS, for which the problem has a global solution, but other definitions of optimal distance preservation often lead to better visualizations. These versions are all non-linear, so both the optimization criterion and the method of optimization can lead to different results.

IsoMap [37] is a more recent method intended for points which lie on a lower-dimensional surface in the high-dimensional space. It begins by constructing a neighborhood graph connecting nearby points in the input space. Then, using this graph, it approximates geodesic distances on the surface. Finally it applies MDS to the matrix of geodesic distances.

While the standard approach in both of these methods is to use a Euclidean target space for the embedding, in the past decades, hyperbolic multidimensional scaling has also been proposed. In a nutshell, the original Euclidean distance in the target space is replaced by the hyperbolic distance:

$$d(z_i, z_j) = 2 \tanh^{-1} \frac{|z_i - z_j|}{|1 - z_i \bar{z_j}|}, \tag{5.2}$$

where z_i and z_j denote two points in the target space. The modification in the distance metric makes the computation of gradients non-trivial. Our goal was to explore the question of whether hyperbolic space would be a more successful target space for the visualization of distributions of trees, since tree space and hypoerbolic space are both non-positive curved.

Recently, Cvetkovski and Crovella [12] introduced a method MDS-PD (metric multidimensional scaling algorithm using the Poincaré disk model) which is based on a steepest decent method with hyperbolic line search. We adapted this software for our experiments with hyperbolic space, and we review the method here; more details can be found in [12]. Complex coordinates are used to present the points of the hyperbolic plane, making the Poincaré disk model a subset of the complex plane $\mathbf{C}$: $\mathbf{D} = \{z \in \mathbf{C} \| z| < 1\}$. The objective function to be minimized is the total embedding error

$$E = c \sum_{j=1}^{n} \sum_{k=j+1}^{n} c_{jk} \left(d_{jk} - \delta_{jk}\right)^2 .$$

where c and c_{jk} are constants, d_{jk} is the hyperbolic distance between points z_j and z_k (Eq. 5.2), and δ_{jk} denotes the dissimilarity/distance between points z_j and z_k in the input dissimilarity/distance matrix. More specifically, we use the Sammon Stress Criterion, in which c and c_{jk} are fixed based on δ_{jk} as follows:

$$E = \frac{1}{\sum_{j=1}^{n} \sum_{k=j+1}^{n} \delta_{jk}} \sum_{j=1}^{n} \sum_{k=j+1}^{n} \frac{\left(d_{jk} - \delta_{jk}\right)^2}{\delta_{jk}} \tag{5.3}$$

This criterion does not favor preserving large distances over small ones. The algorithms starts with a set of random points in the Poincaré disk. In each iteration, it moves each of the points along the gradient direction of the energy function shown in Eq. 5.3 with a Mobius transform until one of the stopping tolerances is met or the maximum iteration number is reached.

5.4.1 Experiments on Real and Synthetic Data

While much of tree-space looks locally like a Euclidean space, there are two local features which are decidedly not Euclidean: corners and open books. A corner is point concentration of negative curvature (see Fig. 5.1b), while an open book is a set of Euclidean half-space attached together along their axes, or "spine" (see Fig. 5.2). These two features, as well as the high dimension of the local Euclidean space, are the sources of error for the low-distortion embedding. We generate synthetic datasets that isolate the two features to determine how hyperbolic MDS (HMDS) and hyperbolic isomap (HIsomap) treat them. We compare the results

Table 5.6 Multiplicative distortion of the embeddings

	MDS	Isomap	HMDS	HIsomap
CORNER	1.4	5.0	18.3	2.96
3SHEETS_2D	71.4	98.9	76.3	44.0
3SHEETS_3D	44.0	189.54	54.1	68.2
5SHEETS_2D	551.9	567.4	87.8	76.1
5SHEETS_3D	2,097.8	470.5	393.6	123.4
COPD_250	253.9	952.3	62.0	64.3

both qualitatively and quantitatively with embeddings done with classical MDS and isomap. More specifically, the datasets are CORNER, in which 250 points are generated by sampling the distance from the origin from a Gaussian distribution $\mathcal{N}(0, 1)$ and sampling an angle with one of the orthant boundaries uniformly from the interval $[0, \frac{5\pi}{2}]$; 3SHEETS_2D, in which 50 points are generated in each of 3 2-dimensional sheets; 3SHEETS_3D, in which 50 points are generated in each of 3 3-dimensional sheets; 5SHEETS_2D, in which 50 points are generated in each of 5 2-dimensional sheets; 5SHEETS_3D, in which 50 points are in each of 5 3-dimensional sheets; and COPD, in which the lung airway trees of 125 healthy patients and 125 patients with COPD are randomly selected. Within each sheet, the 50 points were generated by sampling from a symmetric normal distribution in the underlying Euclidean space that is centered at the origin.

The multiplicative distortion for each embedding approach is summarized in Table 5.6. The multiplicative distortion for a single distance between two points in the dataset is *original_distance/embedded_distance*. The distortion for the whole dataset is *max_distortion/min_distortion*, where *max_distortion* is the maximum distortion of any two points and *min_distortion* is the minimum distortion for any two points. HMDS and HIsomap perform the best for almost all of the datasets. The embedded visualizations and the histograms for each dataset are found in Figs. 5.5 and 5.6.

Qualitatively, for CORNER, all methods were qualitatively able to group the points in the same quadrant, and MDS performs best qualitatively, while the hyperbolic Isomap performs better than the Isomap. For the two-dimensional open books 3SHEETS_2D and 5SHEETS_2D, all methods also grouped the points by their respective sheets. The two Euclidean methods overlaid all but two of the sheets, while the two hyperbolic methods kept the sheets distinct, particularly in 3SHEETS_2D, better representing the true geometry. Despite increasing the dimension only by one, for 3SHEETS_3D it was much harder for the methods to separate the distinct sheets. While MDS performed the best quantitatively, this was not the case qualitatively, where the two hyperbolic methods gave better sheet separation. All methods had trouble representing the more complex datasets 5SHEETS_3D and COPD, although quantitatively, the hyperbolic methods did a far better job of reducing distortion.

The ideal histogram would place all of the distances in the column corresponding to zero error. Although the embedded datasets in Fig. 5.5 do not provide much

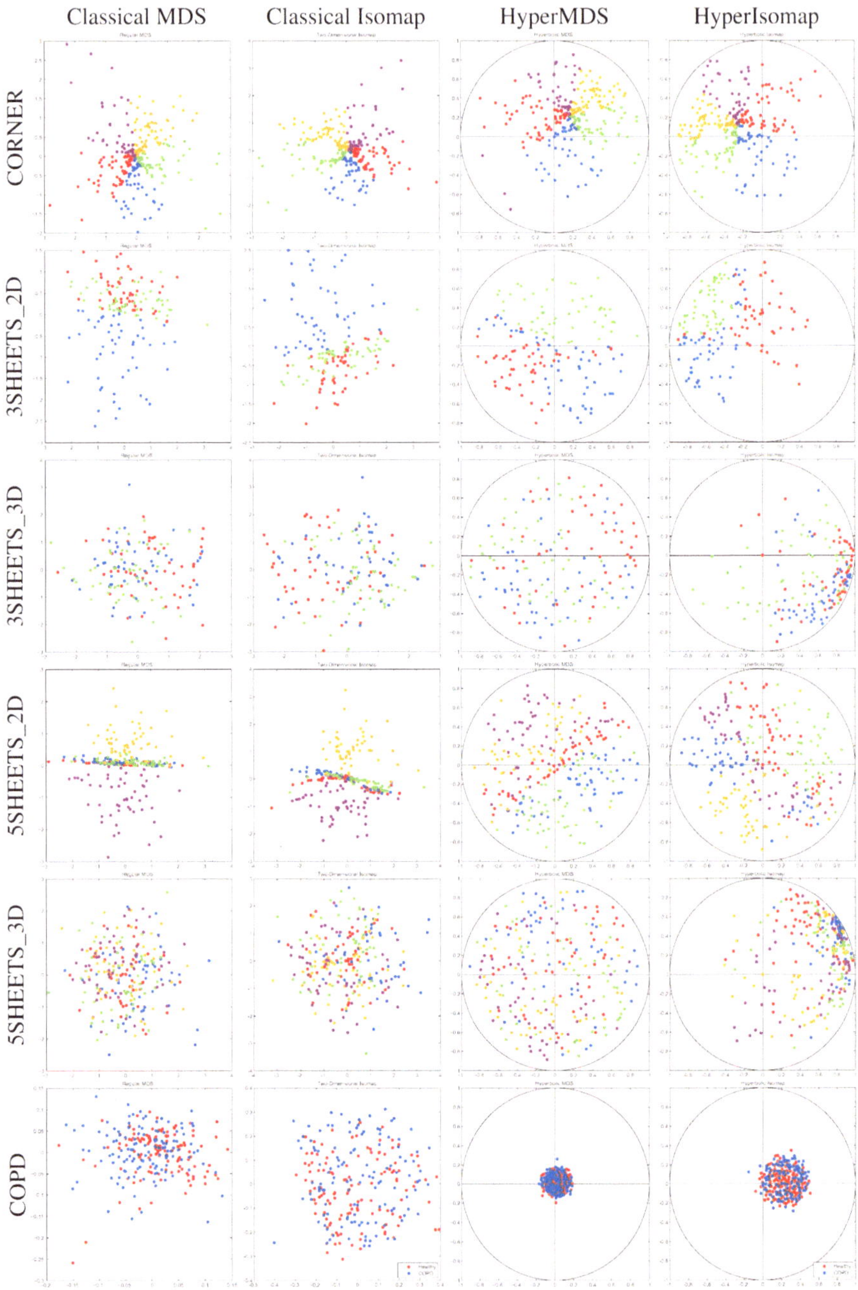

Fig. 5.5 The embedded datasets. For the CORNER, 3SHEETS_2D, 3SHEETS_3D, 5SHEETS_2D, and 5SHEETS_3D dataset embeddings, points have the same color if they are located in the same quadrant or sheet. For the COPD dataset embeddings, the class of healthy patients is colored in *red*, and the class of patients with COPD are colored in *blue*

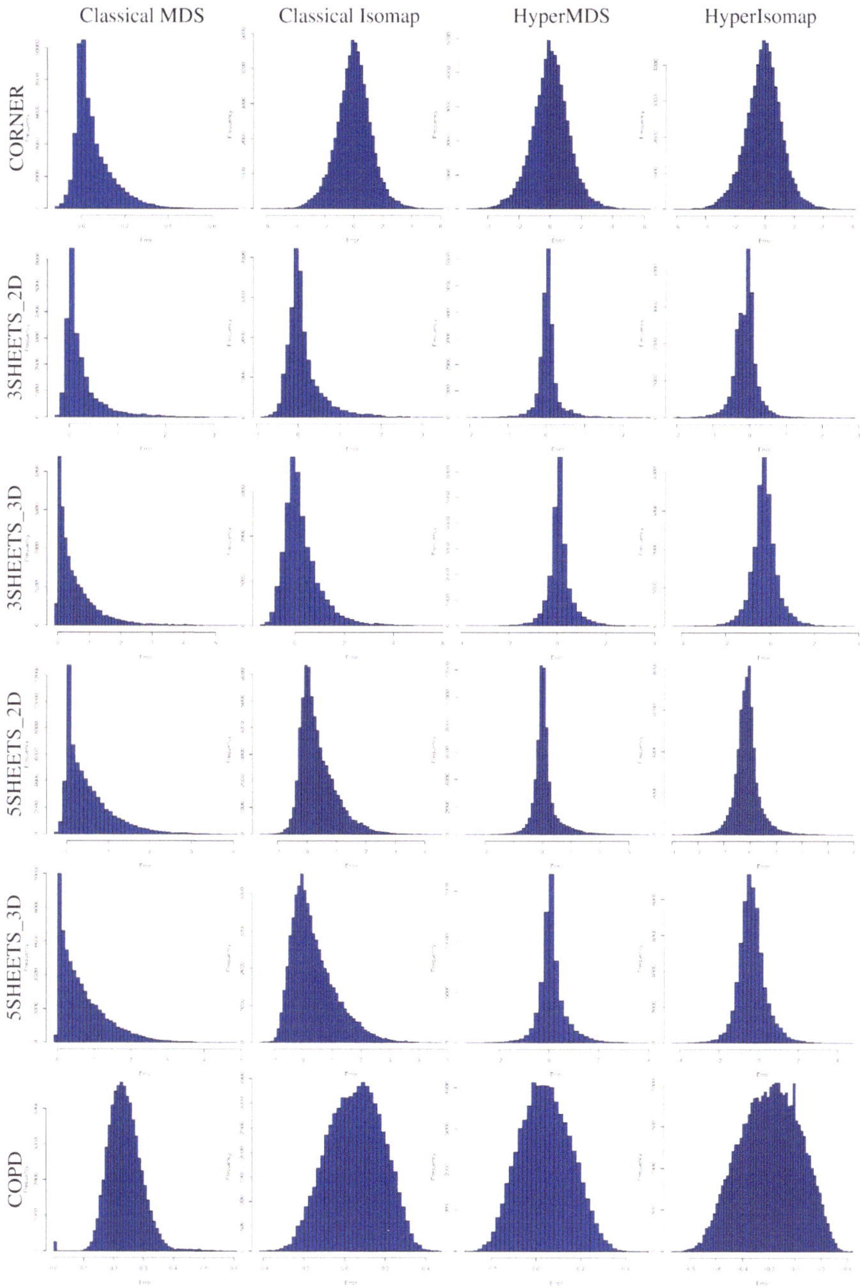

Fig. 5.6 The error histograms of the embedded datasets in Fig. 5.5. Every pair of points is binned according to the error in the embedding, which is the difference between the original distance between the pair of points, and the distance between them in the embedding

qualitative insight for the more complex datasets, the histograms in Fig. 5.6 show that the hyperbolic methods generally give the most accurate reduction to two dimensions.

5.5 Discussion and Conclusion

We have considered two different approaches for quantifying and visualizing variance in datasets of trees. In Sect. 5.3 the dataset trees were divided into nested subtrees, in order to quantify the contribution of different subtrees in distinguishing two populations of trees through either hypothesis testing or classification. These approaches were applied to populations of airway trees from COPD patients and healthy individuals, where the most discriminative subtrees were extracted for the different tasks. In Sect. 5.4 visualization of population structure for datasets of trees was studied through multidimensional scaling and isomap in a hyperbolic disc as opposed to in the Euclidean plane. The choice of a hyperbolic visualization space was motivated by the fact that tree-space itself has singular points which are hyperbolic, and it thus seems likely that a hyperbolic visualization space can give a more truthful rendering of the structure of the population of trees than a Euclidean space. We demonstrate a quantitative and visual improvement in dataset visualization on a set of synthetic datasets sampled from singular spaces representing the types of singularities found in tree-space, as well as on a set of airway trees.

These approaches supply a new set of tools, and give insight into new potential solutions, for analysis of tree-structured data. Future work includes development of structured sparsity methods using subtrees where the correlation between different subtrees is explicitly taken into account, as well as low-distortion embedding into more complex non-Euclidean visualization spaces whose geometry is similar to that of tree-space.

Acknowledgements This work was supported by NSF IIS 111766; MO was supported through a Fields-Ontario Postdoctoral Fellowship; AF was supported by the Danish Council for Independent Research | Technology and Production. MH works for National Security Technologies, LLC, under Contract No. DE-AC52-06NA25946 with the U.S. Department of Energy/National Nuclear Security Administration, DOE/NV/25946--2016.

We thank the organizers, Kathryn Leonard and Luminita Vese, and the sponsors of the collaboration workshop Women in Shape (WiSh): Modeling Boundaries of Objects in 2- and 3-Dimensions that was held at the Institute for Pure and Applied Mathematics (IPAM) at UCLA July 15-19 2013, where the collaboration leading to this paper was established.

References

1. Agarwala, R., Bafna, V., Farach, M., Paterson, M., Thorup, M.: On the approximability of numerical taxonomy (fitting distances by tree metrics). SIAM J. Comput. **28**(3), 1073–1085 (1999)
2. Amenta, N., Klingner, J.: Case study: visualizing sets of evolutionary trees. In: IEEE Symposium on Information Visualization (INFOVIS 2002), Boston, pp. 71–74. IEEE (2002)
3. Ashburner, J., Friston, K.J.: Voxel-based morphometry the methods. Neuroimage **11**, 805–821 (2000)
4. Azencott, C.-A., Grimm, D., Sugiyama, M., Kawahara, Y., Borgwardt, K.M.: Efficient network-guided multi-locus association mapping with graph cuts. Bioinformatics **29**(13), 171–179 (2013)
5. Bacak, M.: Computing medians and means in Hadamard spaces. SIAM J. Optim. **24**(3), 1542–1566 (2014)
6. Badoiu, M., Indyk, P., Sidiropoulos, A.: Approximation algorithms for embedding general metrics into trees. In: SODA, New Orleans, pp. 512–521 (2007)
7. Billera, L.J., Holmes, S.P., Vogtmann, K.: Geometry of the space of phylogenetic trees. Adv. Appl. Math. **27**(4), 733–767 (2001)
8. Borg, I., Groenen, P.: Modern Multidimensional Scaling. Springer, Heidelberg (1997)
9. Busch, J.R., Ferrari, P.A., Flesia, A.G., Fraiman, R., Grynberg, S.P., Leonardi, F.: Testing statistical hypothesis on random trees and applications to the protein classification problem. Ann. Appl. Stat. **3**(2), 542–563 (2009)
10. Cates, J., Fletcher, P.T., Styner, M., Shenton, M., Whitaker, R.: Shape modeling and analysis with entropy-based particle systems. In: Karssemeijer, N., Lelieveldt, B. (eds.) Information Processing in Medical Imaging. Lecture Notes in Computer Science, vol. 4584, pp. 333–345. Springer, Berlin/New York (2007)
11. Chakerian, J., Holmes, S.: Computational tools for evaluating phylogenetic and hierarchical clustering trees. J. Comput. Graph. Stat. **21**(3), 581–599 (2012)
12. Cvetkovski, A., Crovella, M.: Multidimensional scaling in the Poincaré Disk (2011). arXiv:1105.5332[stat.ML]
13. Feragen, A., Kasenburg, N., Petersen J., de Bruijne, M., Borgwardt, K.: Scalable kernels for graphs with continuous attributes. In: NIPS, Lake Tahoe (2013)
14. Feragen, A., Owen, M., Petersen, J., Wille, M.M.W., Thomsen, L.H., Dirksen, A., de Bruijne, M.: Tree-space statistics and approximations for large-scale analysis of anatomical trees. In: IPMI, Asilomar, pp. 74–85 (2013)
15. Feragen, A., Petersen, J., Owen, M., Lo, P., Thomsen, L.H., Wille, M.M.W., Dirksen, A., de Bruijne, M.: A hierarchical scheme for geodesic anatomical labeling of airway trees. In: MICCAI (3), Nice, pp. 147–155 (2012)
16. Friedman, J., Hastie, T., Tibshirani, R.: Regularization paths for generalized linear models via coordinate descent. J. Stat. Softw. **33**(1), 1–22 (2010)
17. Gasparovic, E.: The Blum medial linking structure for multi-region analysis. PhD Dissertation, University of North Carolina at Chapel Hill (2012)
18. Geers, A.J., Larrabide, I., Radaelli, A.G., Bogunovic, H., Kim, M., Gratama van Andel, H.A.F., Majoie, C.B., VanBavel, E., Frangi, A.F.: Patient-specific computational hemodynamics of intracranial aneurysms from 3D rotational angiography and CT angiography: an in vivo reproducibility study. Am. J. Neuroradiol. **32**(3), 581–586 (2011)
19. Ghanbari, Y., Smith, A.R., Schultz, R.T., Verma, R.: Connectivity subnetwork learning for pathology and developmental variations. In: Proceedings of the 16th International Conference Medical Image Computing and Computer-Assisted Intervention (MICCAI 2013), Nagoya, 22–26 Sept 2013, Part I, pp. 90–97 (2013)
20. Gorczowski, K., Styner, M., Jeong, J.-Y., Marron, J.S., Piven, J., Hazlett, H.C., Pizer, S.M., Gerig, G.: Multi-object analysis of volume, pose, and shape using statistical discrimination. IEEE Trans. Pattern Anal. Mach. Intell. **32**(4), 652–661 (2010)

21. Hillis, D.M., Heath, T.A., St John, K.: Analysis and visualization of tree space. Syst. Biol. **54**(3), 471–482 (2005)
22. Huang, J., Zhang, T., Metaxas, D.: Learning with structured sparsity. In: Proceedings of the 26th Annual International Conference on Machine Learning (ICML 2009), Montreal, pp. 417–424 (2009)
23. Jenatton, R., Gramfort, A., Michel, V., Obozinski, G., Eger, E., Bach, F., Thirion, B.: Multiscale mining of fMRI data with hierarchical structured sparsity. SIAM J. Imaging Sci. **5**(3), 835–856 (2012)
24. Lo, P., Sporring, J., Pedersen, J.J.H., Bruijne, M.: Airway tree extraction with locally optimal paths. In: Medical Image Computing and Computer-Assisted Intervention (MICCAI 2009), London. Lecture Notes in Computer Science, vol. 5762, pp. 51–58 (2009)
25. Lo, P., van Ginneken, B., Reinhardt, J.M., Yavarna, T., de Jong, P.A., Irving, B., Fetita, C., Ortner, M., Pinho, R., Sijbers, J., Feuerstein, M., Fabijanska, A., Bauer, C., Beichel, R., Mendoza, C.S., Wiemker, R., Lee, J., Reeves, A.P., Born, S., Weinheimer, O., van Rikxoort, E.M., Tschirren, J., Mori, K., Odry, B., Naidich, D.P., Hartmann, I., Hoffman, E.A., Prokop, M., Pedersen, J.H., de Bruijne, M.: Extraction of airways from CT (EXACT'09). IEEE Trans. Med. Imaging **31**(11), 2093–2107 (2012)
26. Megalooikonomou, V., Barnathan, M., Kontos, D., Bakic, P., Maidment, A.: A representation and classification scheme for tree-like structures in medical images: analyzing the branching pattern of ductal trees in x-ray galactograms. TMI **28**(4), 487–493 (2009)
27. Miller, E., Owen, M., Provan, J.S.: Polyhedral computational geometry for averaging metric phylogenetic trees. Adv. Appl. Math. (2015, Accepted)
28. Montealegre, I., St John, K.: Visualizing restricted landscapes of phylogenetic trees. In: Proceedings of the European Conference for Computational Biology (ECCB 2003), Citeseer (2002)
29. Nye, T.M.W.: Principal components analysis in the space of phylogenetic trees. Ann. Stat. **39**(5), 2716–2739 (2011)
30. Owen, M., Provan, J.S.: A fast algorithm for computing geodesic distances in tree space. ACM/IEEE Trans. Comput. Biol. Bioinf. **8**, 2–13 (2011)
31. Pedersen, J., Ashraf, H., Dirksen, A., Bach, K., Hansen, H., Toennesen, P., Thorsen, H., Brodersen, J., Skov, B., Døssing, M., Mortensen, J., Richter, K., Clementsen, P., Seersholm, N.: The Danish randomized lung cancer CT screening trial – overall design and results of the prevalence round. J. Thorac. Oncol. **4**(5), 608–614 (2009)
32. Perez-Cruz, C., Simon, M., Czéh, B., Flügge, G., Fuchs, E.: Hemispheric differences in basilar dendrites and spines of pyramidal neurons in the rat prelimbic cortex: activity- and stress-induced changes. Eur. J. Neurosci. **29**, 738–747 (2009)
33. Petersen, J., Nielsen, M., Lo, P., Nordenmark, L.H., Pedersen, J.H., Wille, M.M.W., Dirksen, A., de Bruijne, M.: Optimal surface segmentation using flow lines to quantify airway abnormalities in chronic obstructive pulmonary disease. Med. Image Anal. **18**, 531–541 (2014)
34. Robinson, D.F., Foulds, L.R.: Comparison of phylogenetic trees. Math. Biosci. **53**, 131–147 (1981)
35. Sturm, K.-T.: Probability measures on metric spaces of nonpositive curvature. Contemp. Math. **338**, 357–390 (2003)
36. Sundberg, K., Clement, M., Snell, Q.: On the use of cartographic projections in visualizing phylogenetic tree space. Algorithms Mol. Biol. **5**(26) (2010). http://www.almob.org/content/5/1/26
37. Tenenbaum, J.B., de Silva, V., Langford, J.C.: A global geometric framework for nonlinear dimensionality reduction. Science **290**(5500), 2319–2323 (2000)
38. Tibshirani, R.: Regression shrinkage and selection via the lasso. J. R. Stat. Soc. Ser. B **58**, 267–288 (1996)
39. Walter, J.: H-MDS: a new approach for interactive visualization with multidimensional scaling in the hyperbolic space. Inf. Syst. **29**(4), 273–292 (2004)

40. Walter, J., Ritter, H.: On interactive visualization of high-dimensional data using the hyperbolic plane. In: Proceedings of the Eighth ACM SIGKDD International Conference on Knowledge Discovery and Data Mining, Edmonton, pp. 123–132. ACM (2002)
41. Wilgenbusch, J.C., Huang, W., Gallivan, K.A.: The evaluation of dimensionality reduction methods to characterize phylogenetic tree-space (2010). Poster at Evolution 2010. Available at http://www.math.fsu.edu/~whuang2/

Chapter 6
Skeleton-Based Recognition of Shapes in Images via Longest Path Matching

Gulce Bal, Julia Diebold, Erin Wolf Chambers, Ellen Gasparovic, Ruizhen Hu, Kathryn Leonard, Matineh Shaker, and Carola Wenk

Abstract We present a novel image recognition method based on the Blum medial axis that identifies shape information present in unsegmented input images. Inspired by prior work matching from a library using only the longest path in the medial axis, we extract medial axes from shapes with clean contours and seek to recognize these shapes within "no isy" images. Recognition consists of matching longest paths

G. Bal (✉)
Department of Computer Engineering, Middle East Technical University, Ankara, Turkey
e-mail: gulcebal@gmail.com

J. Diebold
Department of Computer Science, Technical University of Munich, Munich, Germany
e-mail: julia.diebold@in.tum.de

E. W. Chambers
Department of Mathematics and Computer Science, Saint Louis University, Saint Louis, MO, USA

Research supported in part by NSF grants CCF-1054779 and IIS-1319573
e-mail: echambe5@slu.edu

E. Gasparovic
Department of Mathematics, Duke University, Durham, NC, USA
e-mail: ellen@math.duke.edu

R. Hu
Department of Mathematics, Zhejiang University, Zhejiang, China
e-mail: ruizhen.hu@gmail.com

K. Leonard
Department of Mathematics, California State University Channel Islands, Camarillo, CA, USA

Research supported in part by NSF grant IIS-0954256
e-mail: kleonard.ci@gmail.com

M. Shaker
Department of Electrical Engineering, Northeastern University, Boston, MA, USA
e-mail: shaker@ece.neu.edu

C. Wenk
Department of Computer Science, Tulane University, New Orleans, LA, USA

Research supported in part by NSF grant CCF-0643597
e-mail: cwenk@tulane.edu

K. Leonard, S. Tari (eds.), *Research in Shape Modeling*, Association for Women in Mathematics Series 1, DOI 10.1007/978-3-319-16348-2_6

from the segmented images into complicated geometric graphs, which are computed via edge detection on the (unsegmented) input images to obtain Voronoi diagrams associated to the edges. We present two approaches: one based on map-matching techniques using the weak Fréchet distance, and one based on a multiscale curve metric after reducing the Voronoi graphs to their minimum spanning trees. This paper serves as a proof of concept for this approach, using images from three shape databases with known segmentability (whale flukes, strawberries, and dancers). Our preliminary results on these images show promise, with both approaches correctly identifying two out of three shapes.

6.1 Introduction

We present a method and proof-of-concept for image recognition based on information extracted from the Blum medial axis. Shape recognition and matching based solely on contour points have been shown to perform weakly in the presence of occlusion, partial data, and noise [4, 13, 21]. Unorganized point sets [5] representing boundaries of shapes are often matched using assignment algorithms for graph matching [10]. Another class of methods which use Hausdorff distance to match the edge maps [13] has the advantage of not requiring correspondences of edge features, but they do not necessarily preserve the integrity of shape parts. Global shape representations which are translation, rotation, or scale invariant such as coefficients of Fourier descriptors [19] may result in incorrect matchings due to noise or occlusion. Historically, approaches based on the medial axis have suffered from its instability and complexity in the presence of noise and pixelation. Our approach is designed to bypass those problems while preserving the strengths of the medial axis as a shape descriptor, including meaningful decomposition into parts and stability despite occlusion. Furthermore, our matching techniques are designed to be near-invariant to Euclidean motions (translation, rotation, and scaling).

While shape recognition based on the medial axis has been well-studied for pre-segmented shapes [25], this project is among the first to perform recognition using the medial axis on an unsegmented unknown image. The basic concept builds on previous work which recognizes objects by matching longest paths in the medial axis, but only in the limited setting where the input is a “nice” shape taken from a particular hand-drawn catalog [3]. Here, we apply a similar philosophy to match shapes in the much more challenging domain where the input is an arbitrary image. As a result, we must apply edge detection and other techniques in order to identify significant shape information present in the image. Additionally, whereas [3] uses both the medial skeleton and the radius function, our current results use only the skeleton because extracting reliable radius information from arbitrary edges in an image presents additional challenges.

Since there is no common frame of reference between shapes from our canonical library of possibilities and our input image, we must match an arbitrary path (the longest path from the canonical image) into a messy geometric graph (the Voronoi diagram of the edges detected from our image). We use two different approaches in this work, one based on map-matching using the weak Fréchet distance and the other based on a multiscale curve matching into the minimum spanning tree of the graph computed from the input image edges.

Our initial results indicate that both matching methods perform reasonably well, clearly matching two of our three initial tests to the correct image. The algorithms are reasonably efficient, although the map-matching approach is more computationally intensive due to the exhaustive set of rotations and transformations that must be tested. Testing on a larger database than our three-object set is required to determine the full power of these methods.

6.2 Background

6.2.1 The Medial Axis

The *medial axis* of an object is the set of points which have more than one closest point on the object's boundary. It was first introduced by Blum as a tool for recognizing shapes in biological images [6]. It is known that the medial axis has the same homotopy type as the original shape [18], and therefore it gives a topologically accurate but simpler representation of the shape of an object. In addition, the geometry of the boundary curve is encoded in the geometry of the medial skeleton and its radius function. The *medial axis transform* is the set of points in the medial axis annotated with the radius of the largest inscribed ball centered at each point. This structure can be used to recover the entirety of the original shape. Applications and algorithms using this structure are numerous; see for example the survey by Leymarie and Kimia and the many other references in [17].

6.2.2 Shape Recognition Using the Medial Axis

One of the main motivations for this work is the fact that medial-axis based structures such as the *shock graph* have had notable success with the problem of image recognition among a large database [22, 23, 25]. Each of these algorithms catalogs a set of canonical shape categories by computing the shock graph (an annotated version of the medial axis) for each of the shape instances. The next step is to read input images and attempt to match the shock graphs of the input images against the library of known shapes. These algorithms are based on dynamic programming, and work efficiently since the shock graph is a tree whenever the input shape is simply connected.

Another line of research motivating our work does not use the entire structure of the medial axis, but instead does the matching strictly based on the longest path in the medial axis and its associated radius function. Bai et al. [3] implemented and tested on a library of shapes containing 56 images total, with 4 objects per shape class [2]. Their approach of removing a shape from the library and testing to get the correct classification resulted in a success rate of 98.2 %. In addition, they implemented and tested their method on a larger dataset [22] with 94.4 % accuracy.

Although this matching is naturally less successful for images with high radial symmetry, they nonetheless successfully match input shapes to the correct class for the vast majority of tested images. This is perhaps surprising, given how much rich information about the medial axis is lost when only considering the single longest path. However, the work has so far been applied only to catalogs of images with hand-drawn, clean contours. In this paper, we apply a related method to recognize a shape contained in an arbitrary (noisy) input image.

6.2.3 Map-Matching

Given a graph G embedded in Euclidean space $\mathbb{R}^d$ (most often $\mathbb{R}^2$) and a polygonal curve γ also embedded in $\mathbb{R}^d$, the map-matching problem asks for the path in G which is closest to γ, generally under some distance measure such as the Fréchet distance or weak Fréchet distance. Recently, this problem has been considered in both theoretical and applied settings due to its utility in GIS applications [1, 7, 8]. In this setting, one often has a trajectory (such as is given by a GPS unit placed in a vehicle) which needs to be matched to the closest path on a known road network, modeled as the graph G.

Our setting is slightly different: although the graphs we work with are extracted from images and thus have embeddings in $\mathbb{R}^2$, our input paths are not embedded in the same frame of reference since the scales and orientations of the arbitrary input images can be quite different from the reference images from the library. This variation is somewhat similar to the notion of a graph isomorphism, but here, our input graphs are geometric graphs rather than arbitrary ones. While fast algorithms for Fréchet distance to a geometric graph have been looked at in some limited settings, such as for trees [12], no one previously has considered the problem where the input path is not given as an embedding into the same frame of reference as the graph G, which adds considerably to the difficulty of the problem.

We perform map-matching via the weak Fréchet distance. Let $\gamma_1, \gamma_2 : [0, 1] \rightarrow \mathbb{R}^2$ be two curves in the plane. The weak Fréchet distance δ_{wF} between them is defined as:

$$\delta_{wF}(\gamma_1, \gamma_2) = \inf_{\alpha_1, \alpha_2 : [0,1] \rightarrow [0,1]} \max_{t \in [0,1]} \|\gamma_1(\alpha_1(t)) - \gamma_2(\alpha_2(t))\|,$$

where α_1 and α_2 range over all continuous reparametrizations with $\alpha_1(0) = \alpha_2(0) = 0$ and $\alpha_1(1) = \alpha_2(1) = 1$, and $||.||$ denotes the Euclidean norm. The weak Fréchet distance is a well-suited distance measure for comparing curves as it takes into account the continuity of the curves. In our setting, we consider the set

$\mathcal{T}$ of translations, rotations, and scalings. And the related map-matching problem that we address is to find for a geometric graph G, a curve γ, and any admissible transformation $T \in \mathcal{T}$, the path in G that minimizes the weak Frechet distance to any $T(\gamma)$.

6.2.4 $H^{1/2}$-Type Multiscale Curve Metric

Our other method of matching an input path into a geometric graph is via the $H^{1/2}$ multiscale curve metric, first introduced in [16], evaluated on a curve extracted from the graph and the known longest medial path. The last decade has produced a substantial body of work on finding shape metrics that respect the underlying geometry of shape space, where a shape is modeled as a curve in $\mathbb{R}^2$ possibly modulo a group of transformations [14, 24, 26]. Unfortunately, these metrics are computationally expensive and can be unwieldy to implement in any realistic setting. The $H^{1/2}$-type metric is a middle-ground: a weakened linearization of a Riemannian metric that is computationally fast. In other words, it computes distances based on geometric quantities whereas the Fréchet distance does not.

For ease of exposition, results here are given for plane curves as objects in $\mathbb{C}$ instead of $\mathbb{R}^2$. We trust the reader can move naturally between these two representations. Given a smooth arclength-parameterized open plane curve $\gamma(s)$, define an $H^{1/2}$ "norm"[1] as:

$$\|\gamma\|_{\frac{1}{2}}^2 = \int_0^L \int_0^{\min(s,L-s)} \beta(s,t)^2 \, dt \, ds,$$

where L is the length of the curve, and the angle $\beta(s,t)$ between the rays joining $\gamma(s)$ to $\gamma(s+t)$ and $\gamma(s-t)$ is given by:

$$\beta(s,t) \equiv \arg \frac{\gamma(s+t) - \gamma(s)}{\gamma(s) - \gamma(s-t)}.$$

Moreover, β gives rise to a metric on curves. Let Σ be the set of homeomorphisms $\sigma : [0,1] \to [0,1]$ and γ_1, γ_2 be Lipschitz curves. Then:

$$L(\gamma_1, \gamma_2) = \inf_{\sigma \in \Sigma} \int \int (\beta_1(s,t) - \beta_2(\sigma(s),t))^2 \, ds \, dt$$

[1] We are not viewing the space of plane curves as linear, but the integral defined is analogous to Sobolev norms on function spaces and the integrand is analogous to a wavelet decomposition of γ. Additionally, the "norm" gives rise to a metric on curves in the standard way.

gives the metric:

$$d^2(\gamma_1,\gamma_2) = L(\gamma_1,\gamma_2) + L(\gamma_2,\gamma_1).$$

For a discretized curve sampled over dyadic intervals, we have:

$$\|\gamma\|^2_{\frac{1}{2}} = \sum_{n=1+2^{k-1}}^{N-2^{k-1}} \sum_{k=1}^{K} \beta(n,k)^2 2^{-k},$$

where N is the number of sampled points, K determines the maximum number of dyadic intervals, and the angle β is:

$$\beta(n,k) = \arg \frac{\gamma(n+2^{k-1}) - \gamma(n)}{\gamma(n) - \gamma(n-2^{k-1})}.$$

If γ is an arclength parameterization of a Lipschitz graph, then the angles $\beta(n,k)$ are, in a distributional sense, the same as the wavelet coefficients of γ over the same dyadic interval system. In this way, the collection of angles $\{\beta(n,k)\}$ provides a multiscale analysis of the curve γ and, in turn, the Haar coefficients of γ' provide a fast computation for $\{\beta(n,k)\}$ based on scaled second differences:

$$\sum_{n=1+2^{k-1}}^{N-2^{k-1}} \sum_{k=1}^{K} \beta(n,k)^2 2^{-k} = \sum_{n=1+2^{k-1}}^{N-2^{k-1}} \sum_{k=1}^{K} \left(\gamma(n+2^k) - 2\gamma(n) + \gamma(n-2^k)\right) 2^{-k}. \tag{6.1}$$

If γ_1 and γ_2 are sampled by $M \le N$ points, respectively, then $\sigma : \{1,\ldots,M\} \to \{1,\ldots,N\}$ and scales are limited by $K \le \log_2 M$ and we obtain the discrete approximation to the continuous metric:

$$L(\gamma_1,\gamma_2) \approx \min_{\sigma\in\Sigma_{M,N}} \sum_{m=1+2^{k-1}}^{M-2^{k-1}} \sum_{k=1}^{K} \frac{1}{k^2} |\beta_1(m,k) - \beta_2(\sigma(m),k)|^2$$

which in turn can be computed using second differences as above.

The metric as defined is naturally translation invariant. In the discrete case, rotation invariance is introduced by rotating the line joining $\gamma(n+2^k)$ and $\gamma(n-2^k)$ to be horizontal (a coarse approximation to the tangent line at $\gamma(n)$) and scale invariance is introduced by normalizing the average inter-point distances to be one. See [16] for details and full generality of results.

6.3 Method

6.3.1 Extracting Medial Axes From "Known" Images

In general, the medial axis of an object in a natural image is difficult to extract automatically, as it requires segmenting the image, extracting the points on the boundary of the object of interest, then computing the medial axis. We select three image databases with known segmentability: whale flukes, strawberries, and dancers. We use k-means clustering to extract an initial binary representation of the object of interest, then apply morphological techniques to obtain a clean boundary. We extract the centers and radii of the circumcircles of the Delaunay triangulation of the boundary points and retain only those centers and radii corresponding to the interior of the object, thereby obtaining the interior medial axis. See Fig. 6.1 for an illustration of this process. For more details on this process, see [15].

To extract the longest path within the axis, we apply Dijkstra's algorithm to find the point P on the axis that is farthest from a randomly selected medial point, then repeat Dijkstra's algorithm to find the medial point Q farthest from P. Retracing steps from Q to P generates the sequence of medial points along the longest path in the medial axis. See Fig. 6.2 for an illustration of this process on our 3 test images.

6.3.2 Extracting Voronoi Edges from "Unknown" Input Images

Given an input image, we smooth it as in [20]. See Fig. 6.3. Let f denote the noisy input image and u the denoised (smooth) version. We obtain u by minimizing the energy:

$$E(u) = \int_{\Omega} (f - u)\, dx + \lambda \int_{\Omega} |\nabla u|\, dx, \tag{6.2}$$

(a) (b) (c)

Fig. 6.1 Intermediate steps for extracting the medial axis from the whale image. The original image can be seen in Fig. 6.6. Images above are (**a**) the initial cluster containing the whale fluke resulting from k-means clustering, (**b**) the segmented whale fluke after morphological processing, and (**c**) the resulting boundary points. The medial axis with longest path resulting from the boundary displayed here can be seen in Fig. 6.2e

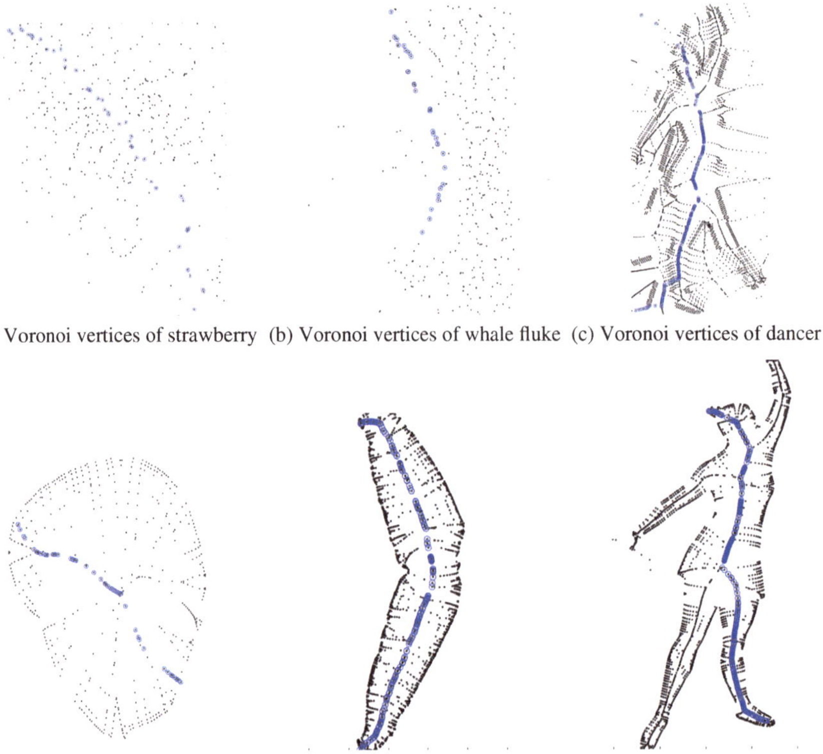

(a) Voronoi vertices of strawberry (b) Voronoi vertices of whale fluke (c) Voronoi vertices of dancer

(d) Medial skeletons of strawberry (e) Medial skeletons of whalefluke (f) Medial skeletons of dancer

Fig. 6.2 *Top row*: Voronoi vertices with longest path highlighted. *Bottom row*: Medial skeletons with longest path highlighted. These longest paths are shown matched to one another in Figs. 6.12–6.14

where Ω denotes the image domain and $\lambda \in \mathbb{R}_{>0}$ a weighting factor. The first term ensures that u is similar to f and the second term forces u to be smooth everywhere except at strong edges.

Next, we run a line segment detector (LSD) algorithm [11] on the smoothed version in order to extract prominent edges and thus a likely boundary of a shape. LSD locally detects straight contours on the image, giving subpixel results while controlling the number of false detections per pixel. Contours are naturally defined by the image gradient and level lines of the image which divide the transition region from dark to light or the opposite. The algorithm works by finding the unit vectors tangent to the level lines, thus computing the level line angle at each pixel. The resulting vector field is then segmented into connected regions that share the same level line angle up to a threshold. Each connected region is represented by a geometrical object such as a rectangle. The principal axis of this object defines the main direction which is chosen as the line segment.

Fig. 6.3 Examples of smoothed images

Fig. 6.4 Results of edge detection on smoothed images of a giraffe and a bottle

The output is a set of edges with noise, as in Fig. 6.4, which we process into a Voronoi diagram to extract potential medial points. In doing so, we remove "outlier" medial points (including points in the region external to the shape) by a dilation and erosion process, as depicted in Fig. 6.5. That is, we first thicken the medial points to form many connected point clusters and subsequently erode them (while still maintaining connected structures). We then identify and delete all point clusters in the processed image of a sufficiently small area, and/or those points that are greater than a certain small distance away from the largest connected structures in the image. We then compare the resulting image with the original input image and delete all medial points in the input image corresponding to deleted points in the processed image, yielding the desired image without outliers.

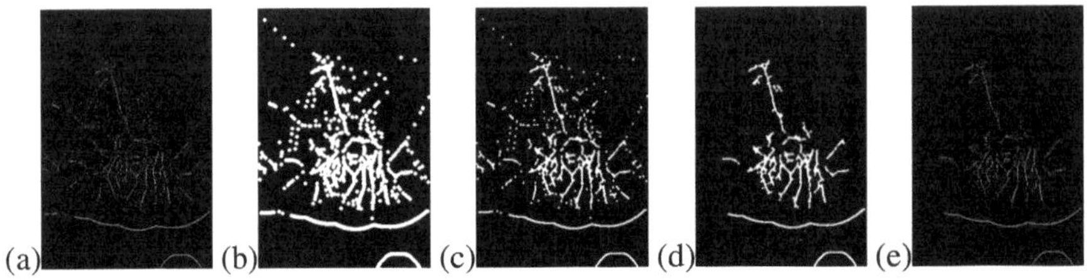

Fig. 6.5 Removing "outlier" medial points for giraffe image. (**a**) Original image with all medial points, followed by the resulting images after (**b**) dilation, (**c**) erosion, and (**d**) point deletion. (**e**) Final image with remaining medial points

Next, our objective is to match the single longest path from each of our initial image instances into the graph which, in each case, approximates the medial axis of the shape that is present in each input image. We pursue this problem in two different ways as outlined in the following two subsections.

6.3.3 Matching via Weak Fréchet Distance

Our first method of matching is based on map-matching via the weak Fréchet distance. The related map-matching problem is to find for a geometric graph G and a curve γ a path in the graph that minimizes the weak Fréchet distance to γ. For a polygonal curve γ with n vertices and a graph G with a total number of m edges and vertices, the map-matching problem can be solved in $O(mn \log(mn))$ time [27]. This algorithm constructs a "free space graph" which is essentially a combinatorial representation of the product space of (parameterizations of) the curve and the graph. Each vertex-edge pair is assigned a weight that equals their Euclidean distance, and then a shortest path algorithm in this "free space graph" (where the length of the path is computed as the maximum of the weights) computes a path with minimum weak Fréchet distance. Please see [27] for more details.

In our setting, we consider the set $\mathcal{T}$ of translations, rotations, and scalings. And the related map-matching problem that we address is to find for a geometric graph G, a curve γ, and any admissible transformation $T \in \mathcal{T}$, the path in G that minimizes the weak Frechet distance to any $T(\gamma)$. We sample $\mathcal{T}$ by applying a fairly exhaustive set of scalings, translations, and rotations to the curve, and for each such transformation we run the map-matching algorithm of Wenk et al. [27]. In particular, we sample the transformation space as follows: We consider rotations by 0°, 90°, 180°, and 270°. We hold the aspect ratio constant and apply a single scaling factor; the maximum scaling factor is determined such that the width of the (possibly rotated) path equals the width of the graph, and the minimum scaling factor is chosen to be half the maximum factor; this range is sampled in steps of 0.2. The two-dimensional translation space is determined to consist of all translations such that the bounding box of the (possibly scaled and rotated) path fits entirely inside the bounding box of the graph; the translation space is sampled in steps of 10 pixels. As described in Sect. 6.4, the dimensions of each bounding box are several hundred pixels by several hundred pixels. The resulting range of scales where between 1 and 2.4 for the strawberry, between 0.38 and 0.78 for the whale fluke, and between 0.6 and 1.2 for the dancer. We note that this method is computationally intensive for each example, involving multiple tests for different possible orientations and sizes of the path.

6.3.4 Matching via an $H^{1/2}$-Type Metric

Our second method of matching addresses the fact that the Fréchet based algorithm described in Sect. 6.3.3 is especially difficult because the input medial axis graph can be quite noisy and messy depending on how well our edge detection and smoothing algorithms are able to isolate prominent shapes. Additionally, the second method applies a metric that is invariant under Euclidean motion.

We first simplify the Voronoi graph to a tree to avoid cycles when computing the longest path in the graph. We choose the minimum spanning tree because it appears to capture the prominent shape features quite well, though other ways of simplifying the input graph may be worth investigating. Note that in converting the graph to a tree we may lose segments on the longest path. Suppose γ_1 is a discrete representation of the longest path in the medial axis of a known object, and we are given the Voronoi edges from an unknown image. Our method is as follows, with curve matching running in $O(MN \log M)$ time:

1. Compute the minimal spanning tree for the Voronoi edges.
2. Extract γ_2, the longest path in the Voronoi tree.
3. Resample γ_1 and γ_2 to have $N = M = 128$ equally spaced points.
4. Normalize scale so that each curve has an average inter-point distance of one.
5. Compute second differences as described in Eq. 6.1.
6. Extract second differences corresponding to every fourth point on γ_1 to allow for flexible point matching (otherwise the points are matched one-to-one in order), following the procedure outlined in Sect. 6.2.4.
7. Apply dynamic programming to find the matching of points of γ_1 to γ_2 that minimizes the distance $d^2(\gamma_1, \gamma_2)$ between the curves.
8. Sum scaled second differences corresponding to the optimal matching to obtain approximation to $d^2(\gamma_1, \gamma_2)$.

6.4 Results

The three images we use are of a strawberry, a whale fluke, and a dancer. Here we match the medial axis extracted as described in Sect. 6.3.1 to the Voronoi diagram of the same image extracted as described in Sect. 6.3.2. The dimensions of the

Fig. 6.6 Input images: a strawberry, a whale fluke, and a dancer

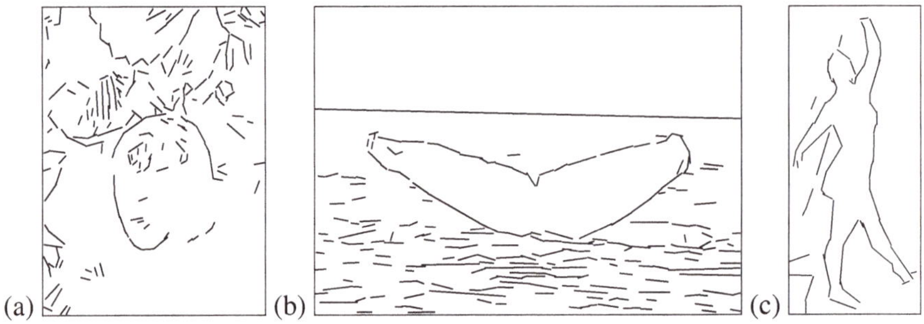

Fig. 6.7 Results of edge detection: (**a**) strawberry, (**b**) whale fluke, and (**c**) dancer

bounding boxes of the Voronoi diagrams are 479×367 for the strawberry, 618×418 for the whale fluke, and 540×239 for the dancer. Results from the two matching methods are comparable, and both seem promising.

6.4.1 Weak Fréchet Map-Matching Distance Results

For the dancer and the whale fluke, the transformation that minimized the weak Fréchet distance over all sampled transformations was found correctly, see Figs. 6.8 and 6.9. The point matching computed by the weak Fréchet distance also appears to be of good quality. The distance for the minimum transformation computed for the dancer is so small (2.5 pixels), that the transformed dancer path and the resulting matched path in the graph almost coincide. For the whale fluke, the minimum computed transformation (11.2 pixels) is very close to the transformation with the third smallest distance of 11.6 pixels which applies an additional 180° transformation to the whale path. For the strawberry, the path is found for multiple

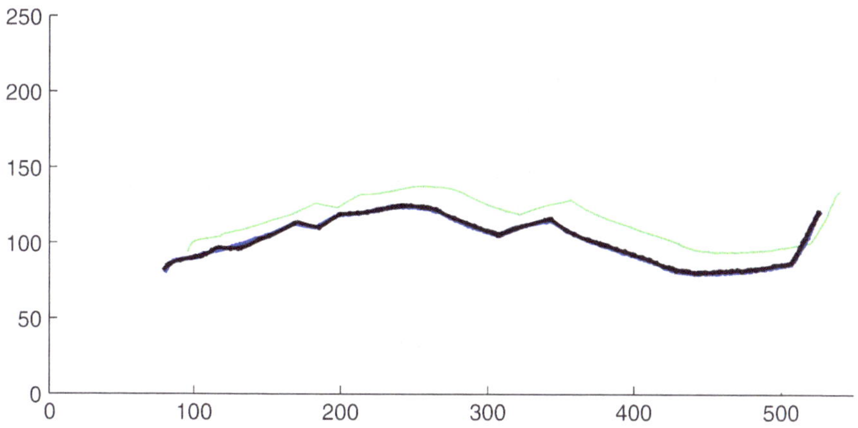

Fig. 6.8 Matching the dancer path into the dancer graph. The graph edges are shown in *light gray*, and the path is shown in *green*. The algorithm finds the correct transformation with minimum Fréchet distance 2.5 pixels (at scale 1.0 with no rotation). The transformed path is shown in *black*, and the corresponding path in the graph in *blue*

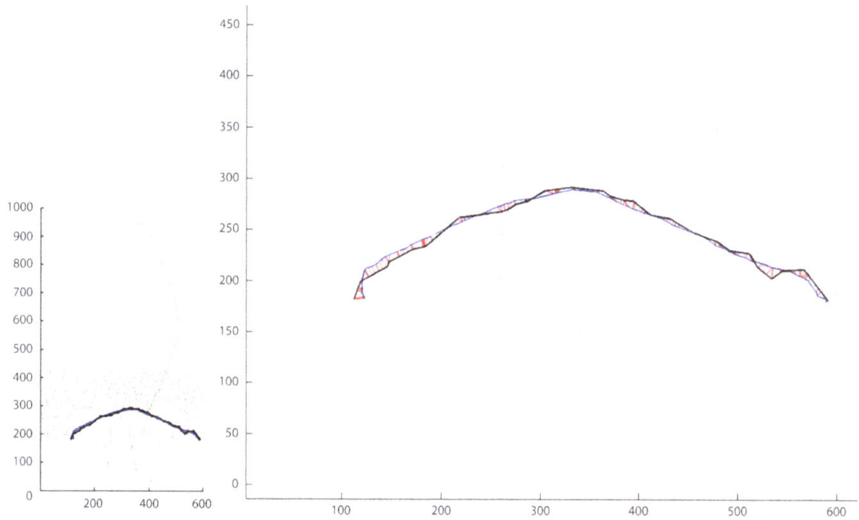

Fig. 6.9 Matching the whale path into the whale graph. The graph edges are shown in *light gray*, and the path is shown in *green*. The algorithm finds the correct transformation with minimum Fréchet distance 11.2 pixels (at scale 0.6 with 90° rotation). The transformed path is shown in *black*, and the corresponding path in the graph is *blue*. The *red lines* show the optimal point matching

small scales at multiple positions at small distances (ranging between 8.1 to about 13) in the graph, see Fig. 6.8. Out of the 4,368 sampled transformations per sampled scale, 9.2 % of the transformations at scale 1.0 have a distance less than 15. At scale 1.2, this reduces to 5.5 %, and at scale 2.0 this reduces to only 0.07 %.

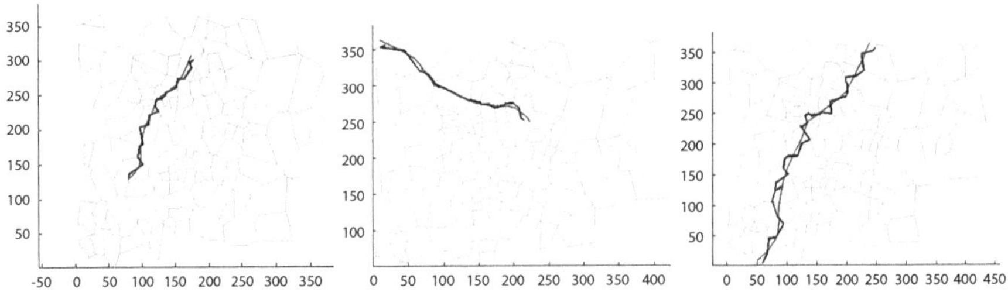

Fig. 6.10 Matching the strawberry path into the strawberry graph. The algorithm finds too many occurrences of the path at a small scale. The graph edges are shown in *light gray*, and the path is shown in *green*. Results are shown for the minimum distance (8.1) at scale 1.0 (with a rotation of 180°), the minimum distance (8.9) at scale 1.2 (with a rotation of 270°), and the minimum distance (13.4) at scale 2.0 (with a rotation of 180°). The transformed path is shown in *black*, and the corresponding path in the graph in *blue*

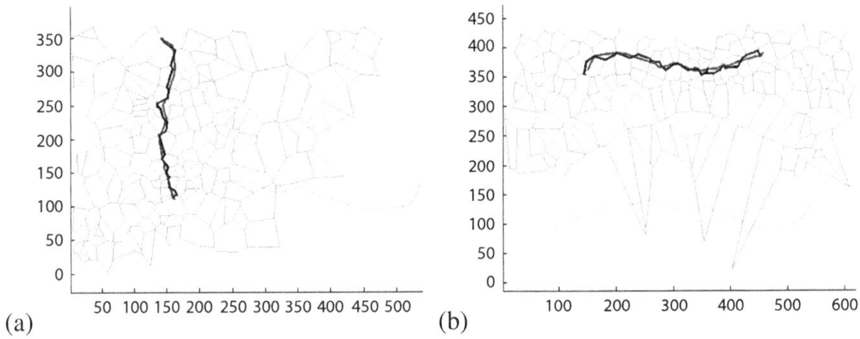

Fig. 6.11 Matching the dancer path into the strawberry graph (distance 8.9) and into the whale fluke graph (distance 8.7). Both distances are larger than the distance into the dancer graph (2.5), see Fig. 6.8

We believe that this is an artifact caused by the almost grid-like dense edge pattern in the strawberry graph in combination with the very straight shape of the strawberry path.

We also compared the dancer path to the strawberry graph, the whale fluke graph, and the dancer graph. We computed the minimum weak Fréchet distance over all sampled transformations. The computed minimum distances were 8.9 pixels for the strawberry graph, 9.7 pixels for the whale fluke graph, and 2.5 for the dancer graph. The dancer path therefore correctly determined the dancer graph as the graph it matches best with, see Figs. 6.8 and 6.11.

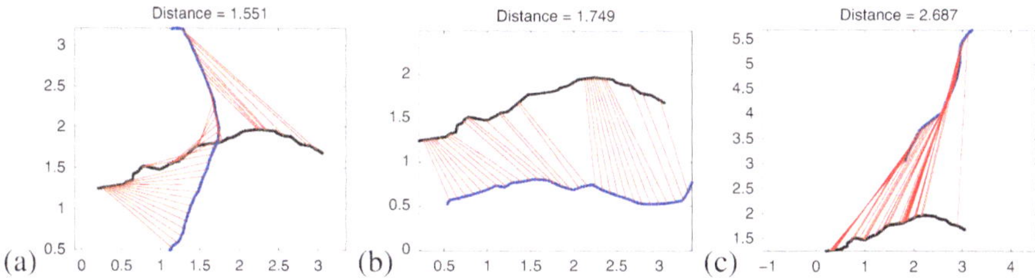

Fig. 6.12 Matching medial longest path from (**a**) whale (distance = 1.551), (**b**) dancer (distance = 1.749), (**c**) berry (distance = 2.687) into the Voronoi tree longest path of the whale fluke. Lines show the optimal point matching. The minimum distance into the messy graph correctly classifies the unknown image as a whale

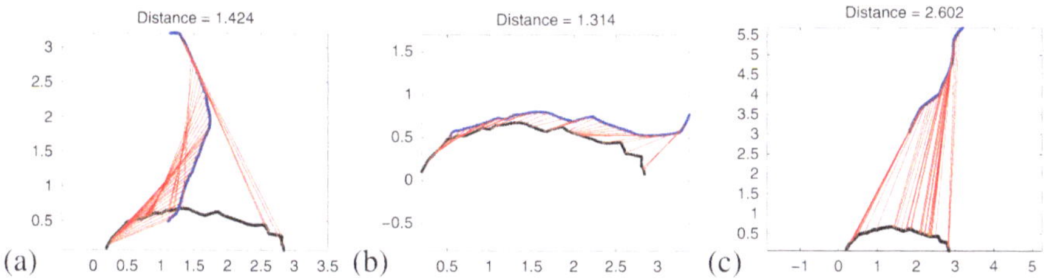

Fig. 6.13 Matching medial longest path from (**a**) whale (distance = 1.424), (**b**) dancer (distance = 1.314), (**c**) berry (distance = 2.602) into the Voronoi tree longest path of the dancer. Lines show the optimal point matching. The minimum distance into the messy graph correctly classifies the unknown image as a dancer

6.4.2 $H^{1/2}$ *Metric Results*

Initial results for matching the medial longest path to the Voronoi tree longest path are correct for the two instances where the longest Voronoi path contains the desired medial points. Apart from the strawberry image, where the Voronoi tree longest path fails to contain edges belonging to the medial axis of the strawberry, the closest match corresponds to the correct classification. In addition, the optimal matching between points performs reasonably well. See Figs. 6.12–6.14. Note that the scale of the curves has changed. This is because of the scale invariance we introduced by normalizing inter-point distances to be one.

6.5 Discussion and Future Directions

Our matching techniques show enough promise to merit additional investigation. We are curious about the success of the algorithms when the Voronoi diagram is dense or grid-like, or where the longest path in the medial axis does not trace a prominent shape feature (such as when the input image is nearly round with radial symmetry).

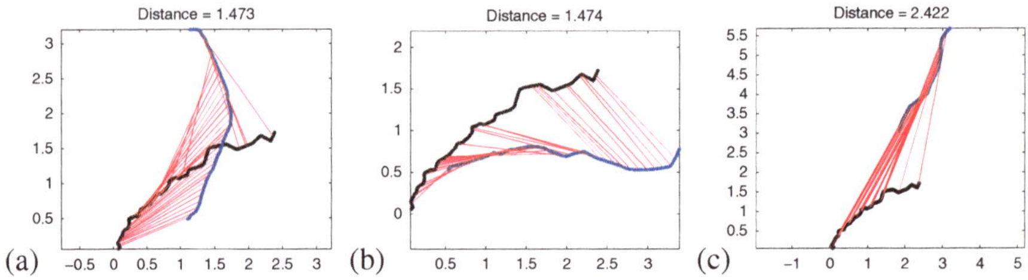

Fig. 6.14 Matching medial longest path from (**a**) whale (distance = 1.473), (**b**) dancer (distance = 1.474), (**c**) berry (distance = 2.422) into the Voronoi tree longest path of the strawberry. Lines show the optimal point matching. The minimum distance into the messy graph incorrectly classifies the unknown image as a whale. Note that the longest path in the Voronoi tree for the strawberry image does not contain any edges from the medial axis of the strawberry itself

6.5.1 Analysis of the Weak Fréchet Map-Matching Distance

Sampling the transformation space to minimize the weak Fréchet map-matching works well for the dancer and the whale fluke. The strawberry graph exhibits a grid-like dense edge pattern which causes the strawberry path to be found in many locations in the graph, in particular for small scales. While this behavior is extreme in the strawberry, it is also present in the whale fluke data, where the path with the second smallest distance is located at a different location with an additional 180° rotation. We believe that the "small scale" problem could be overcome by analyzing the distribution of distances for fixed scale and varying translations and rotations, in order to identify transformations with significant distances. We will investigate this direction in future research.

For the dancer path, the distance into the dancer graph was much smaller than into the strawberry graph and the whale fluke graph. The minimum weak Fréchet distance into the messy graph therefore correctly classifies the unknown image as a dancer.

6.5.2 Analysis of the $H^{1/2}$-Type Metric

Matching longest paths using the $H^{1/2}$-type metric performs well for the two cases, whale and dancer, where the longest path in the Voronoi tree contains edges corresponding to the medial axis of the object of interest. Not surprisingly, it fails for the third image, the strawberry, where no medial edges appear in the Voronoi tree longest path. The strawberry image is particularly challenging, as the berry itself contributes very few edges to the very complicated edge map seen in Fig. 6.7a and contains several spurious edges in its interior. This illustrates the need for an additional evaluation of relative importance of Voronoi vertices, perhaps through classification of vertices as belonging to the foreground or background or noise.

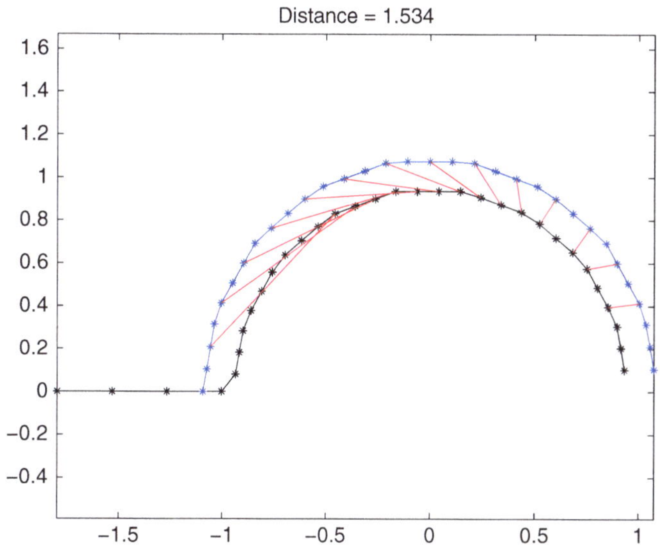

Fig. 6.15 Matching a semicircle into a curve composed of the union of a semicircle and a line. Instead of matching semicircle to semicircle, the semicircle is matched to points away from the line

In addition, the optimal matching of points between the two longest paths currently seems to favor matches that map the medial path into the length of the Voronoi path. For example, in Fig. 6.12 the medial axis for the whale in the Voronoi path starts at about the halfway point whereas the optimal matching begins at the left of the path. Because the optimal matching can skip enough points to avoid highly mismatching segments, it seems likely that a longer match will often be lower cost for medial curve matching. At the same time, two curves that are identical up to a point can correspond in a match that is too short. Figure 6.15 illustrates this issue. For the larger scales, the differences in the β angles (and their second difference approximations) will grow as the points on the circles approach the points on the line. Hence the lowest cost match avoids points toward the end of the circle attached to the line. Penalizing skips that are longer than an average skip, or adding the difference in the radius function values to the cost of matching two points may improve the medial point correspondence.

6.5.3 Future Work

Our initial proof of concept for this approach is promising. Based on our results and prior work in this area [3], we speculate that this approach will also work well to capture the same shape in a different pose (such as a dancer in different positions). Future work will consider a larger library of shapes as well as input images in different poses. There is also potential to include the radius function as well as the longest path for improved recognition results, but it is not clear if the Voronoi graphs from input images will prove too noisy to reliably calculate this information.

Both approaches perform better with a simpler Voronoi graph. We are currently exploring methods for evaluating the saliency of either a particular Voronoi vertex or (equivalently) an edge pair associated to a Voronoi vertex. In addition, both approaches would benefit from using the information in the radius function on the medial and Voronoi points that gives the distance to the corresponding edge points. We anticipate substantial improvement from the combination of these modifications.

We also hope to reduce the cost of learning additional shape classes once a sufficient number of classes have been learned. Learning the visual models for classification of test objects requires a significant number of training samples. In the method of one-shot learning [9], the information from previously learned categories is used for training new categories, using a Bayesian prior and maximum a posteriori (MAP) estimation. This model could be used to optimize and extend learning for the current methods.

Acknowledgements The authors would like to thank the Institute for Pure and Applied Mathematics, the Association for Women in Mathematics, Microsoft Research, the National Science Foundation, and the National Geospatial Agency for support, financial and otherwise, of this collaboration. Kathryn Leonard thanks Matt Feiszli for providing the initial Matlab code for the $H^{1/2}$ metric for closed curves which was modified for this project.

References

1. Alt, H., Efrat, A., Rote, G., Wenk, C.: Matching planar maps. J. Algorithms **49**(2), 262–283 (2003)
2. Aslan, C., Tari, S.: An axis-based representation for recognition. In: ICCV, Beijing, pp. 1339–1346. IEEE Computer Society (2005)
3. Bai, X., Yang, X., Yu, D., Latecki, L.J.: Skeleton-based shape classification using path similarity. Int. J. Pattern Recognit. Artif. Intell. (IJPRAI) **22**(4), 733–746 (2008)
4. Belongie, S., Malik, J., Puzicha, J.: Shape matching and object recognition using shape contexts. IEEE Trans. Pattern Anal. Mach. Intell. **24**(4), 509–522 (2002)
5. Belongie, S., Mori, G., Malik, J.: Matching with shape contexts. In: Analysis and Statistics of Shapes, pp. 81–105. Birkhäuser (2005)
6. Blum, H.: A transformation for extracting new descriptors of shape. In: Wathen-Dunn, W. (ed.) Models for the Perception of Speech and Visual Form, pp. 362–380. MIT, Cambridge (1967). http://citeseer.nj.nec.com/context/77000/0
7. Brakatsoulas, S., Pfoser, D., Salas, R., Wenk, C.: On map-matching vehicle tracking data. In: Proceedings of the 31st International Conference on Very Large Data Bases (VLDB'05), Trondheim, pp. 853–864. VLDB Endowment (2005)
8. Chen, D., Driemel, A., Guibas, L.J., Nguyen, A., Wenk, C.: Approximate map matching with respect to the Fréchet distance. In: Müller-Hannemann, M., Werneck, R.F.F. (eds.) ALENEX, San Francisco, pp. 75–83. SIAM (2011)
9. Fei-Fei, L., Fergus, R., Perona, P.: One-shot learning of object categories. IEEE Trans. Pattern Anal. Mach. Intell. **28**(4), 594–611 (2006)

10. Gold, S., Rangarajan, A.: A graduated assignment algorithm for graph matching. IEEE Trans. Pattern Anal. Mach. Intell. **18**(4), 377–388 (1996)
11. Grompone von Gioi, R., Jakubowicz, J., Morel, J.-M., Randall, G.: LSD: a Line Segment Detector. Image Processing On Line (2012)
12. Gudmundsson, J., Smid, M.: Fréchet queries in geometric trees. In: Bodlaender, H.L., Italiano, G.F. (eds.) Algorithms – ESA 2013, Sophia Antipolis. Lecture Notes in Computer Science, vol. 8125, pp. 565–576. Springer, Berlin/Heidelberg (2013)
13. Huttenlocher, D.P., Klanderman, G.A., Rucklidge, W.J.: Comparing images using the Hausdorff distance. IEEE Trans. Pattern Anal. Mach. Intell. **15**(9), 850–863 (1993)
14. Kushnarev, S.: Teichons: solitonlike geodesics on universal Teichmüller space. Exp. Math. **18**, 325–336 (2009)
15. Leonard, K., Strawbridge, R., Lindsay, D., Barata, R., Dawson, M., Averion, L.: Minimal geometric representation and strawberry stem detection. In: 2013 13th International Conference on Computational Science and Its Applications (ICCSA), Ho Chi Minh City, pp. 144–149 (2013)
16. Leonard, K., Feiszli, M., Kushnarev, S.: Metric spaces of shapes and applications: compression, curve matching and low-dimensional representation. Geom. Imaging Comput. **1**(2), 173–221 (2014)
17. Leymarie, F.F., Kimia, B.B.: From the infinitely large to the infinitely small: applications of medial symmetry representations of shape. In: Siddiqi, K., Pizer, S.M. (eds.) Medial Representations: Mathematics, Algorithms and Applications. Computational Imaging and Vision, pp. 327–351. Kluwer Academic, Dordrecht (2008). ISBN:978-1-402-08657-1. http://opac.inria.fr/record=b1130467
18. Lieutier, A.: Any open bounded subset of $\mathbb{R}^n$ has the same homotopy type as its medial axis. Comput.-Aided Des. **36**(11), 1029–1046 (2004)
19. Lin, C.C., Chellappa, R.: Classification of partial 2-D shapes using Fourier descriptors. IEEE Trans. Pattern Anal. Mach. Intell. **5**, 686–690 (1987)
20. Rudin, L.I., Osher, S., Fatemi, E.: Nonlinear total variation based noise removal algorithms. Physica D **60**, 259–268 (1992)
21. Sebastian, T.B., Kimia, B.B.: Curves vs. skeletons in object recognition. Signal Process. **85**(2), 247–263 (2005)
22. Sebastian, T.B., Klein, P.N., Kimia, B.B.: Shock-based indexing into large shape databases. In: Proceedings of the 7th European Conference on Computer Vision-Part III (ECCV'02), Copenhagen, pp. 731–746. Springer, London (2002)
23. Sebastian, T.B., Klein, P.N., Kimia, B.B.: Recognition of shapes by editing their shock graphs. IEEE Trans. Pattern Anal. Mach. Intell. **26**(5), 550–571 (2004)
24. Sharon, E., Mumford, D.: 2D-shape analysis using conformal mapping. Int. J. Comput. Vis. **70**, 55–75 (2006)
25. Trinh, N.H., Kimia, B.B.: Skeleton search: category-specific object recognition and segmentation using a skeletal shape model. Int. J. Comput. Vis. **94**(2), 215–240 (2011)
26. Trouve, A., Miller, M.I., Younes, L.: On metrics and Euler-Lagrange equations of computational anatomy. Ann. Rev. Biomed. Eng. **4**, 375–405 (2002)
27. Wenk, C., Salas, R., Pfoser, D.: Addressing the need for map-matching speed: localizing global curve-matching algorithms. In: Proceedings of the 18th International Conference on Scientific and Statistical Database Management (SSDBM'06), Vienna, pp. 379–388. IEEE Computer Society, Washington, DC (2006)

Chapter 7
Revisiting Skeletons from Natural Images

Erkut Erdem and Sibel Tari

Abstract In the last two decades there have been several works promoting shape fields that implicitly encode local convexity/concavity properties of the shape boundary. These shape fields are formulated either as solutions to Poisson type PDEs or via heuristic approximations to them. The v-field of Tari-Shah-Pien, can be computed directly from a real image; thus, suggests a mechanism to bridge low level visual processing and high level shape computations. We revisit Tari, Shah and Pien's v-field approach and extend its application to complex images with texture. We relate v-field value at a skeleton point to the distance of the point from a putative shape boundary, and use this relation to extract semantic image patches. At the end of the chapter, we experimentally compare the medial locus computed from the new v-field to that of Kimia et al.

7.1 Introduction

Real valued functions defined on the shape interior is a classical approach for implicitly coding shape characteristics, with motivations reaching back to Gestalt School [8]. There are several ways of defining these encoding maps. The most common of such mappings is the distance transform which assigns each point in the shape interior the points minimum distance to the boundary. During the last two decades more informative variants implicitly coding boundary curvature have been developed [1, 2, 4, 9, 12], all of which are formulated either as a solution to the screened Poisson PDE or heuristic approximations to it.

E. Erdem
Department of Computer Engineering, Middle East Technical University, Ankara, Turkey

Present address: Department of Computer Engineering, Hacettepe University, Ankara, Turkey
e-mail: erkut@cs.hacettepe.edu.tr

S. Tari (✉)
Department of Computer Engineering, Middle East Technical University, Ankara, Turkey
e-mail: stari@metu.edu.tr

K. Leonard, S. Tari (eds.), *Research in Shape Modeling*, Association for Women in Mathematics Series 1, DOI 10.1007/978-3-319-16348-2_7

One potentially useful feature of the v field [9, 10] is that it can be computed directly from a real image as the steady-state solution to a coupled set of PDEs:

$$\frac{\partial u}{\partial t} = \nabla \cdot (v^2 \nabla u) - \frac{\beta}{\alpha}(u - f); \quad \left.\frac{\partial u}{\partial n}\right|_{\partial \Omega} = 0\,, \tag{7.1}$$

$$\frac{\partial v}{\partial t} = \nabla^2 v - \frac{2\alpha |\nabla u|^2 v}{\rho} - \frac{(v-1)}{\rho^2}; \quad \left.\frac{\partial v}{\partial n}\right|_{\partial \Omega} = 0 \tag{7.2}$$

where $f : \Omega \rightarrow \Re$ is a given image defined on $\Omega \subset \Re^2$ of which boundary is $\partial \Omega$, and α, β, ρ are the scale space parameters. The coupled set of equations define an evolution for a smoothed image u and a function v. Notice that if u is fixed, (7.2) becomes the minimiser of

$$\int_{\Omega} \left(\frac{\rho^2}{1 + 2\alpha\rho|\nabla u|^2} |\nabla v|^2 + \left(v - \frac{1}{1 + 2\alpha\rho|\nabla u|^2} \right)^2 \right) d\mathbf{x} \tag{7.3}$$

This means that v is a blurred form of $\frac{1}{1+2\alpha\rho|\nabla u|^2}$ where ρ is the blur radius. As $\rho \rightarrow 0$, $1 - v$ can be thought as the probability of an edge. As ρ increases, however, v starts to code the minimum distances to putative edges.

If the input image f is piecewise constant then v is the minimiser of

$$\int_{\Omega} \rho^2 |\nabla v|^2 + (v-1)^2 \, d\mathbf{x} \tag{7.4}$$

subject to homogeneous Dirichlet conditions on the jump set. Thus, for silhouettes, v or $1 - v$ is governed by the so-called *Screened Poisson Equation*. There is an interesting relation between the level curves of v and evolution of fronts with curvature dependent speed, which has been proven in [9]: Let r denote the arc-length along the gradient lines of v and $curv(\mathbf{x})$ the curvature of the level curve of v passing through the point $\mathbf{x}$ at $\mathbf{x}$, then for a change δv in the v value, one has to move δr along the gradient lines, where δr is

$$\delta r(\mathbf{x}) \approx \frac{\rho}{1 - v} \left(1 + \frac{\rho\, curv(\mathbf{x})}{2} \right) \delta v(\mathbf{x}) \tag{7.5}$$

Thus, one can imagine a two-parameter family of level curves parameterized by and v and ρ. Smoothing increases with increasing ρ and v.

Based on the above connection, a robust ridge (skeleton) detection scheme has been proposed in [9, 10] via zero-crossings of $\frac{d^2|\nabla v|}{ds^2} > 0$. A sample result is shown in Fig. 7.1 for a cat silhouette for two choices of ρ. The method also provides a distinction between branches tracking protrusions (loci of points where $\frac{d^2|\nabla v|}{ds^2} > 0$, where s is the arc length along the level curves) and those tracking indentations (loci of points where $\frac{d^2|\nabla v|}{ds^2} < 0$).

Fig. 7.1 Illustrative skeletons using the method in [9, 10]. The *top row* shows the level curves of v. The *red marks* the critical locations where $\nabla v = 0$. The *bottom row* depicts the skeletons extracted

7.1.1 A Modified v

In principle, v, hence, shape skeletons, should be computable directly from a raw image using the coupled PDEs (7.1) and (7.2) followed by the ridge extraction scheme of Tari-Shah-Pien [10]. However, this is not practical due to texture, noise and other possible complications. In this chapter, we revisit the approach in [10] so that it can be developed towards a better tool to integrate low level and high level visual processing, *purely in a bottom-up* fashion. To such end, we replace the coupled PDEs with the following pair from [3]:

$$\frac{\partial u}{\partial t} = \nabla \cdot ((cv)^2 \nabla u) - \frac{\beta}{\alpha}(u - f); \quad \left.\frac{\partial u}{\partial n}\right|_{\partial \Omega} = 0\,, \tag{7.6}$$

$$\frac{\partial v}{\partial t} = \nabla^2 v - \frac{2\alpha |\nabla u|^2 v}{\rho} - \frac{(v-1)}{\rho^2}; \quad \left.\frac{\partial v}{\partial n}\right|_{\partial \Omega} = 0\,. \tag{7.7}$$

The difference between the former and the new coupled PDE set is the way the image u is smoothed. Notice that the only difference is the replacement of the diffusivity v in the update equation of u with cv. The new diffusivity function cv, which steers the smoothing of u (hence indirectly affecting v), introduces high-level influences into the diffusion process by considering image features computed in a wider context. The strength of these features of interests are denoted by the function ϕ, which is explicitly estimated at each image point, and is used to adjust the edge diffusivity as follows:

$$cv = \phi v + (1 - \phi)V . \tag{7.8}$$

If we have a measure ϕ which attains higher values for meaningful occurrences of a feature, we set $V = 1$ to diffuse faster at those points with lower ϕ, but if ϕ attains lower values for meaningful occurrences, we set $V = 0$ to slow down the diffusion where ϕ is low. That is, the type of an influence is determined by the scalar V, which is set either as 0 for boosting the feature of interest or 1 for eliminating it. If $\phi = 1$, hence $c = 1$ at all image points, the set of PDEs is reduced to (7.1) and (7.2). This is nothing but an approximation to [5].

7.2 From Images to Skeleton Pieces

Following the spirit in [10], we will use the same computational machinery for both edge localization and medialness computations. The key point is to change ρ from very low to very high. In Fig. 7.2, a sample case where four different v functions obtained by varying ρ parameter are depicted. The localization of edges coded by the function v deteriorates as ρ increases from 1 to 32. Notice that for $\rho = 32$, the edge localization is very poor, and the v function resembles a distance function. The value of the parameter ρ mainly determines the growth rate of the approximate distance function. The rate of growth increases with the decreasing ρ. As a result, v function encodes the medialness information more reliably. It is important to note that just like the function in [10], the new v function is an implicit function of the image gradient rather than the distance to shape boundary.

The overall scheme is illustrated in Fig. 7.3. The coupled PDE set is employed twice in a row. In the first phase, ρ is set to a very low value to compute a good edge-indicator v and a cartoon-like u. The rest of the scale space parameters are selected

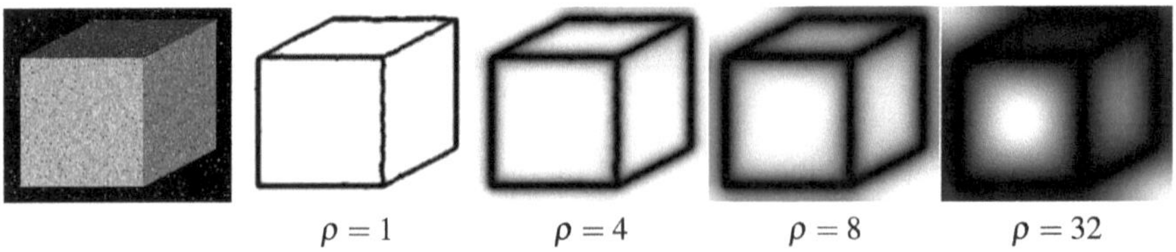

Fig. 7.2 A cube image degraded by Gaussian noise with zero mean and non-constant variance along with 2 % impulse noise and the v function computed via coupled equations using different ρ values. As ρ increases, the behaviour changes from accurately locating the edges to coding medial properties

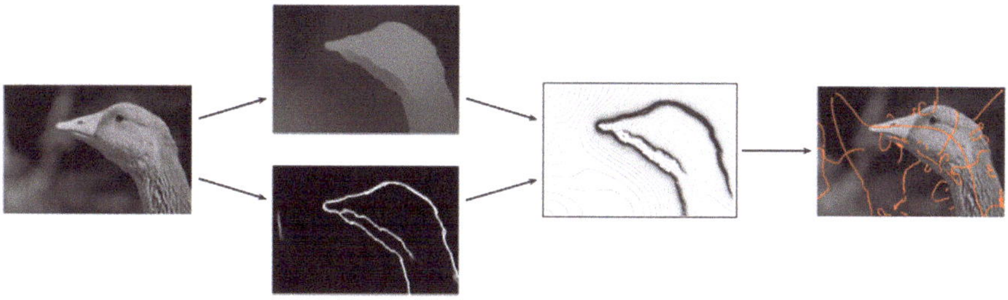

Fig. 7.3 From images to skeleton pieces

based on the intended smoothing of the image (the smoothing radius is $\sqrt{\frac{2\alpha}{\beta}}$ and the contrast threshold is $\sqrt{\frac{1}{2\alpha\rho}}$). The higher level influences are also invoked via c [3]. For this first phase, the coupled PDE set (7.6) and (7.7), with high-level influences in effect allows us to consider the visual content of the input image u in a better way. The result u is a structure-preserving smoothing of the image f. The ambiguities in the visual information are resolved during the joint smoothing process, which *eliminates the need for any preprocessing* to obtain accurate object boundaries. The outcome of the first phase is shown in the second column in Fig. 7.3.

Upon convergence, the parameters are reset: The value of ρ is significantly increased, higher-level influences are shut down by setting $c = 1$ everywhere, and the smoothing of u is shut down by letting $\sqrt{\frac{2\alpha}{\beta}} \to 0$ without increasing the contrast threshold. In the second phase, u does not change much. The v on the other hand changes significantly as shown in the third column via its level curves. The fourth column depicts the skeleton loci extracted from the final v via the method of Tari-Shah-Pien [10].

A nice characteristics of this diffusivity formulation is that it allows one to consider several features of interests at once since their effects are formulated as a multiplier for the edge diffusivity v. For the first phase (the edge localization phase) we compute cv by employing three different feature strength measures ϕ^{dc}, ϕ^{h}, ϕ^{te}, from [3], which respectively denote directional consistency, edge continuity, and texturedness. Such a choice, as detailed in [3], provides a structure-preserving smoothing of f, which allows us to preserve global shapes while eliminating noise, fine details and texture. Results for four more images are given in Fig. 7.4. Recall that the loci can be split into two groups. From now on, our interest is restricted to protrusion branches.

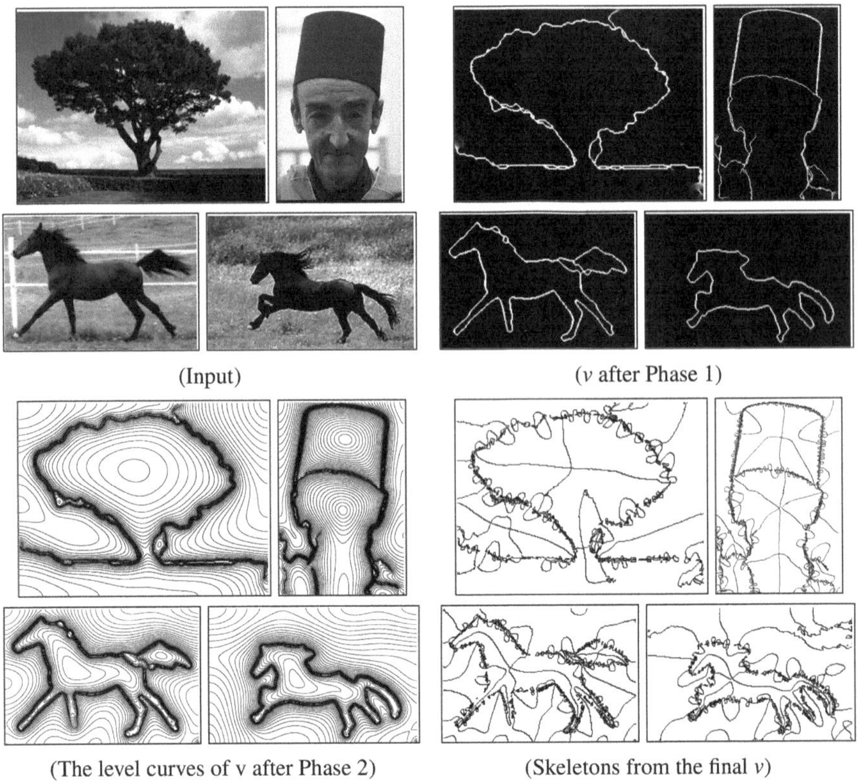

Fig. 7.4 Four illustrative examples for "from real images to skeletons in 3 steps". The skeleton loci is obtained from the final v using the method of Tari-Shah-Pien [10] without any pre or post processing

7.3 From Skeletons to Image Patches

The value of v at a point depends on the gradient $|\nabla u|$ at the point and its neighbors. (Recall that v function is a nonlinearly smoothed form of $\frac{1}{1+2\alpha\rho|\nabla u|^2}$) As a result, one can not estimate the distance from a skeleton point to the nearest boundary, namely the *skeleton radius*, using the value of v. Difficulties associated using the v function for extracting boundaries has been previously reported in [7]. As a solution, we propose to replace the direct dependency of v to $|\nabla u|$, just at a skeleton point, with an indirect dependency through an absolute distance $|d|$ such that the function v attains its minimum at a distance $|d|$ from the skeleton point.

We base our formulation on a ribbon-like section of a shape, i.e. a section having a slowly varying width, and the skeleton points in that shape section. In Fig. 7.5, we provide a graphical illustration, which shows a ribbon-like shape section with the corresponding skeleton points (the dotted line). We consider a cross section over this part, which is illustrated as a red line in the figure, and concentrate on a

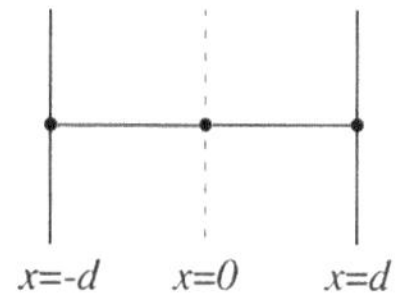

Fig. 7.5 An illustration of a ribbon-like section of a shape and its skeleton (the *dotted line*)

one-dimensional analysis. Note that the boundaries at $|d|$ distance from the skeleton point are the putative edges. Ignoring the curvature effect, the evolution equation is reduced to:

$$\left(\frac{d^2}{dx^2} - \frac{2\alpha|u_x|^2}{\rho} - \frac{1}{\rho^2}\right) v = -\frac{1}{\rho^2}; \quad -d \le x \le d \tag{7.9}$$

with the conditions $v(-d) = v(d) \approx 0$.

In Eq. (7.9), $|u_x|$ is large at the putative edges, which makes $\frac{1}{1+2\alpha\rho|\nabla u|^2}$ have being practically zero at $x = \pm d$. On the other hand, $|u_x|$ is a small quantity unless $v \approx 0$. Therefore, it can be argued that the second and third terms in the left hand side are negligible. Additionally, the term in the right hand side affects the scaling of the solution, i.e. multiplying the right hand side by ρ^2 does not make a qualitative difference. Following these arguments, it can be claimed that the behavior is roughly governed by

$$\frac{d^2 v}{dx^2} = -\frac{1}{\rho^2}; \quad -d \le x \le d\ . \tag{7.10}$$

Consequently,

$$v(0) = \frac{d^2}{2\rho^2}\ . \tag{7.11}$$

Using Eq. (7.11), the radius r of the maximal circle is given by

$$r = d = \rho\sqrt{2v}\ . \tag{7.12}$$

Note that it is also possible to consider, that is to ignore only the middle term in (7.9) without considering the case $\rho \to \infty$:

$$\left(\frac{d^2}{dx^2} - \frac{1}{\rho^2}\right) v = -\frac{1}{\rho^2}; \quad -d \le x \le d \tag{7.13}$$

In that case,

$$1 - v(x) = \left(\frac{1 - e^{\frac{2d}{\rho}}}{e^{\frac{-2d}{\rho}} - e^{\frac{2d}{\rho}}}\right) e^{\frac{-x}{\rho}} - \left(\frac{1 - e^{\frac{-2d}{\rho}}}{e^{\frac{-2d}{\rho}} - e^{\frac{2d}{\rho}}}\right) e^{\frac{x}{\rho}}$$

Thus,

$$r = d = \rho\, cosh^{-1}\left(\frac{1}{1-v}\right) \tag{7.14}$$

In practice, we do not see much difference between the results obtained by (7.14) and (7.12). Hence, we will use (7.12) as it is computationally less expensive.

In Figs. 7.6 and 7.7, we illustrate sample skeletal circles with the radii computed using Eq. (7.12). Notice that the envelopes of the circles are quite in agreement with the corresponding ribbon-like shape sections. But also note that when the shape parts deviate from the ribbon-like structure, the resulting circles become slightly larger and we lose some accuracy in the estimated distances to the putative edges. As the part we consider deviates from ribbon-like, e.g., ear of an animal, radii are overestimated. In practice, we may consider a conservative radius by ignoring $\sqrt{2}$. The regularization inherent to our formulation is the main reason of these partially

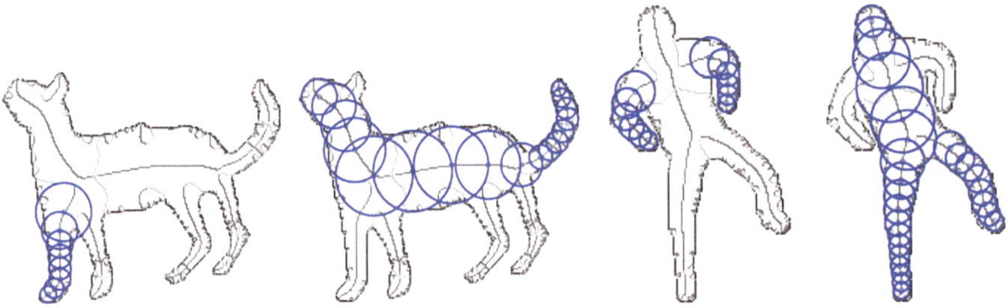

Fig. 7.6 Sample skeletal circles with radii computed using Eq. (7.12)

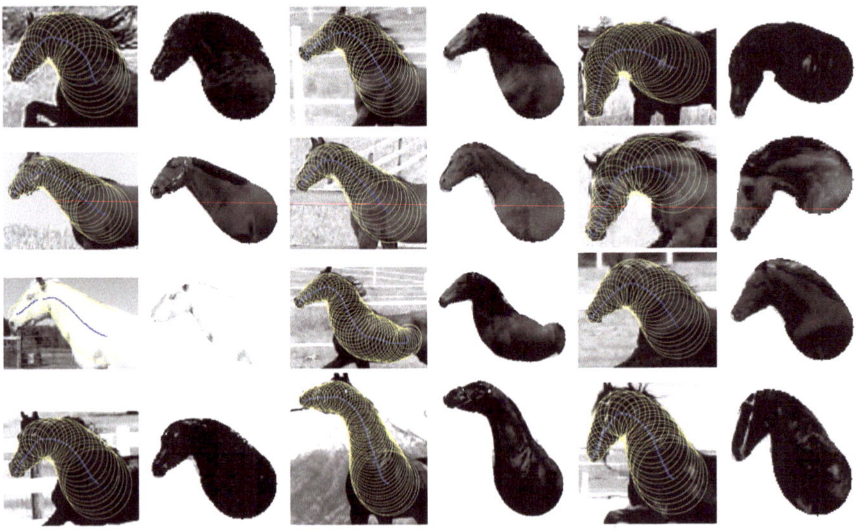

Fig. 7.7 Circles associated with the skeleton points on selected branches for 12 horses

inaccurate radii values, however it is also crucial for obtaining more coarse skeletons (see Fig. 7.1). Unlike the distance transform that accurately encodes every detail in the boundary via a one-to-one mapping, which provides a precise measure of the minimum distance of the point to the shape boundary, our medialness function encodes only the visually meaningful parts and ignores the secondary and unimportant details.

7.4 Experimental Results

In this section we present two groups of experiments. In the first group, Figs. 7.8 and 7.9, we compare skeletons via our method to the ones by Kimia method [6, 11]. In the second group, Figs. 7.10 and 7.11, we depict how patches obtained by the envelope of circles can be employed for figure-ground separation.

7.4.1 Experimental Comparison to Kimia's Method

Our method, being linear, has enormous computational advantages over the method of Kimia. This has been discussed extensively in [9, 10]. Here, we only do compare resulting medial loci of the two respective approaches visually. We have chosen

(Kimia's shocks [11])

(Tari-Shah-Pien skeletons with modified v-field [3].)

Fig. 7.8 Our framework copes well with texture and produces visually more meaningful skeletons (the *bottom row*) whereas Kimia's method produces superfluous shocks and even on textureless areas (the body of the horse or the bird) can not yield medial loci

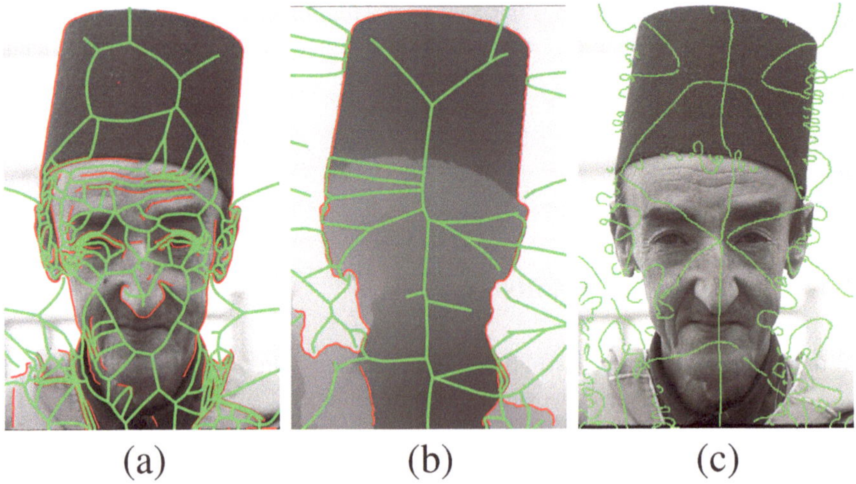

Fig. 7.9 (**a**) Kimia's method applied to raw input. (**b**) Kimia's method applied to our u-result. (**c**) Tari-Shah-Pien method applied on our u-result

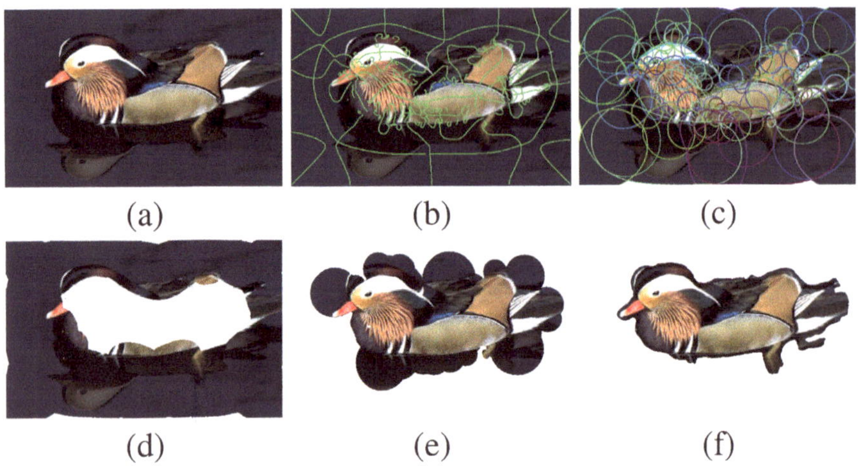

Fig. 7.10 An illustrative segmentation. (**a**) Input image. (**b**) Extracted skeletons. (**c**) Structured image patches. (**d**)–(**e**) Background/foreground regions estimated by the first clustering phase taking extra input from the user. (**f**) The refined separation result

three illustrative real images: an image containing a highly textured cheetah on a uniform background, an image containing low textured horse on a textured background , and a low-textured bird image (Figs. 7.8).

Since there are highly textured regions in the input images, the edge maps of the images contain gaps and many spurious edges. As a consequence, Kimia's method produces many inaccurate and unintuitive shocks loci. (The contour fragments where the shocks are extracted are shown in red colour in the top row images.)

Fig. 7.11 Input images with Tari-Shah-Pien skeletons superimposed on them (the *top row*) and resulting segmented figure regions (the *bottom row*). The suggested figure-ground segmentation algorithm gives fairly good results even though simple texture features are used to describe the structured image patches

The advantage of our method can be attributed to several factors. But one of the factors is the robust structure preserving smoothing achieved via (7.6) [3]. This step is independent of the loci extraction part. Therefore, in our next experiment, to make a direct comparison between the two loci extraction methods given respectively in [11] and [10], we apply Kimia's shock extraction method on our u-images. In Fig. 7.9, we depict three results: (a) Kimia's method applied on a raw input; (b) Kimia's method applied on a denoised and de-textured image (structure preserving smoothing applied to the raw image via (7.6)); and (c) Tari-Shah-Pien method applied on the same denoised and de-textured image. As expected, the shocks obtained from our u-images (in the middle) are much more intuitive than the shocks obtained from raw images (on the left), but not as intuitive as the medial loci via Tari-Shah-Pien method (on the right). Observe that, when Tari-Shah-Pien method is used (on the right), unintuitive medial branches due to inaccuracies near face or fez boundary terminate without corrupting the main branches capturing the two rectangular areas.

7.4.2 *Application to Figure-Ground Separation*

Specifically, using extracted skeleton branches as input (Fig. 7.10b), we extract patches from highly textured images and form them into coherent groups to obtain semantic segments. The presented segmentation algorithm involves the following steps: First, the structured image patches, which are of uniform characteristics and which encode specific sections of the image, are determined (Fig. 7.10c).

Each structured patch is then described by means of simple texture features such as mean intensity, standard deviation, entropy, homogeneity of the co-occurrence matrix. Following that, a k-means clustering with $k = 5$ is carried out on these features. As the input images all contain objects which are distinguishable from the surroundings by means of intensity or texture difference, at least one of the resulting clusters includes the regions that correspond to background. At this point, the labels (figure/background) are manually assigned to the clusters by taking an input from the user. It is important to note that this is the only part of our algorithm where the user intervention occurs. As can be seen in Fig. 7.10d–e, this intermediate step provides a rough, but fairly good separation of the foreground from the background. Now that a rough foreground/background separation is available, a thorough analysis is performed to refine the separation result. The main idea of this step is to restrict the attention to the initially estimated foreground portion of the image during re-estimation of the separation. For that, first, the convex hull enclosing the initial foreground region is estimated. Each pixel in the convex hull is then described by means of the entropy value calculated within a squared window centering the pixel (ignoring the pixels outside the convex hull), and an additional clustering (k-means with $k = 2$) is performed on these entropy values. As a result, the pixels are classified into two as foreground and background, yielding a more accurate figure-ground segmentation than the initial separation result (Fig. 7.10f).

Figure 7.11 shows some extra test images and the corresponding figure-ground segmentations. Although the foreground objects (birds and butterflies) have large variations in texture, pose and scale patterns and simple texture features are used to describe our structured image patches, the proposed algorithm gives fairly good outcomes.

Acknowledgements We thank O. Ozcanli for running Kimia method [6] on our data. The work reported here is initiated under the grant TUBITAK 105E154 and completed with financial support of grant TUBITAK 112E208.

References

1. Aslan, C., Tari, S.: An axis-based representation for recognition. In: ICCV, Beijing, pp. 1339–1346 (2005)
2. Bergbauer, J., Tari, S.: Wimmelbild analysis with approximate curvature coding distance images. In: SSVM, Graz, pp. 489–500 (2013)
3. Erdem, E., Tari, S.: Mumford-Shah regularizer with contextual feedback. J. Math. Imaging Vis. **33**(1), 67–84 (2009)
4. Gurumoorthy, K., Rangarajan, A.: A Schroedinger equation for the fast computation of approximate Euclidean distance functions. In: SSVM, Voss, pp. 100–111 (2009)
5. Mumford, D., Shah, J.: Optimal approximations by piecewise smooth functions and associated variational problems. Commun. Pure Appl. Math. **42**(5), 577–685 (1989)
6. Ozcanli, O.C., Kimia, B.B.: Generic object recognition via shock patch fragments. In: BMVC, Warwick, pp. 1030–1039 (2007)

7. Shah, J.: A common framework for curve evolution, segmentation and anisotropic diffusion. In: CVPR, San Francisco, pp. 136–142 (1996)
8. Tari, S.: Fluctuating distance fields, parts, three-partite skeletons. In: Innovations for Shape Analysis, pp. 439–466
9. Tari, S., Shah, J., Pien, H.: A computationally efficient shape analysis via level sets. In: Mathematical Methos in Biomedical Image Analysis, pp. 234–243 (1996) IEEE Computer Society Press, Los Alamitos, California (2013)
10. Tari, S., Shah, J., Pien, H.: Extraction of shape skeletons from grayscale images. Comput. Vis. Image Underst. **66**(2), 133–146 (1997)
11. Tek, H., Kimia, B.B.: Symmetry maps of free-form curve segments via wave propagation. Int. J. Comput. Vis. **54**(1–3), 35–81 (2003)
12. Zucker, S.: Distance images and the enclosure field. In: Innovations for Shape Analysis, pp. 301–323. Springer, Berlin/New York (2013)

Chapter 8
Towards Automated Filtering of the Medial Axis Using the Scale Axis Transform

Jeannine Abiva and Lisa J. Larsson

Abstract This paper analyzes the problem of determining the optimal scaling to prune the medial axis of spurious branches with the use of the Scale Axis Transform (SAT) in $\mathbb{R}^2$. This optimal scaling is found by minimizing the Fréchet distance between the boundary of the true shape and the boundary of the SAT-filtered version of the shape perturbed by noise. To compute the minimum, the noisy shape is filtered using a variety of scalings $s > 1$ of the SAT algorithm. The optimal scaling is then related to the level of noise used to perturb the true shape. The minimization problem is repeated for various shapes and different noise levels. In applications such as image recognition and registration, the medial axis is very relevant. However, it is highly susceptible to noise along the boundary. The results presented here offer crucial information to automate the de-noising process, by providing a link between the level of noise and the optimal SAT scaling factor.

8.1 Introduction

The medial axis is an important tool in geometry processing as it is homotopy-equivalent to the shape itself [14], and can be more efficient for shape modeling [12, 13]. As such, it has found many applications, including shape recognition [15, 21, 22, 24]. When dealing with shapes whose boundaries have been distorted by noise, it is necessary to prune spurious parts of the medial axis in a robust way. There have been a variety of approaches to address this problem [2, 3, 6, 15, 23]. A recent algorithm that produces pruned skeletons is the Scale Axis Transform

J. Abiva (✉)
Department of Mathematics, The University of Iowa, 14 MacLean Hall,
Iowa City, IA 52242, USA
e-mail: jeannine.abiva@gmail.com

L.J. Larsson
Department of Mathematics and Statistics, McGill University, 805 Sherbrooke Street West,
Montreal, QC H3A 0B9, Canada
e-mail: lisa.powers@mail.mcgill.ca

K. Leonard, S. Tari (eds.), *Research in Shape Modeling*, Association
for Women in Mathematics Series 1, DOI 10.1007/978-3-319-16348-2_8

(SAT) [9, 10, 17]. For any scale $s < 1$ and a class of scales greater than one, topologically-consistent medial axes are produced [10]. Under this algorithm, the medial axis is computed, then the radii of the medial balls are scaled by a factor s to produce a new, enlarged shape. The medial axis of the new shape is computed, and the radii of the medial balls are then scaled back by $1/s$ to obtain the simplified skeleton.

When applying the SAT algorithm to de-noise the medial axes of a database of shapes – for instance during the recognition process – it is not practical to find the scale parameter s manually for each shape. Rather, a best guess should be posited based on the level of the noise of the shape boundary relative to the size of the shape. Previous work involving the SAT algorithm has demonstrated the effect of different scales on the resulting medial axis with a view towards pruning [9, 10, 19]. In this paper, we extend this idea by finding the best scaling to remove spurious branches of the medial axis, which in turn will reconstruct a shape perturbed by noise. Using a database of non-noisy shapes, we perturb them by a discrete set of realistic noise levels. Noise on the shape boundary can be obtained, for example, by computing the shape boundary from a pixelated image [20]. Therefore, given a noisy shape, we call the SAT parameter s optimal if it minimizes the Fréchet distance between the original and SAT-filtered shape. The process by which we compute the optimal scaling is summarized in Fig. 8.1. By investigating the changes to the boundary and the medial axis under different scale factors, we can assess how the topology of the medial axis and the geometry of the shape change under each scaling and noise regime.

The organization of this paper is as follows. In Sect. 8.2, we describe the previous work that is used to identify optimal scale factors. This includes the details of the Scale Axis Transform, previous definitions of distance between shapes, the realistic boundary noise model we adopt, as well as a description of the software package to visualize the results of the SAT algorithm and the medial axis. In Sect. 8.3, we describe the method used to arrive at an optimal SAT scale factor. Section 8.4 then presents the results of this methodology on different input shapes, and Sect. 8.5 follows with concluding remarks.

8.2 Previous Work

In this section, the Scale Axis Transform, as well as previous work on boundary noise and shape distances, is described. This is followed by a summary of the software package Mesecina, which allows for easy visualization of 2D and 3D shape geometries, including the medial axis and the Scale Axis Transform. Now, we begin by describing the SAT algorithm in more detail.

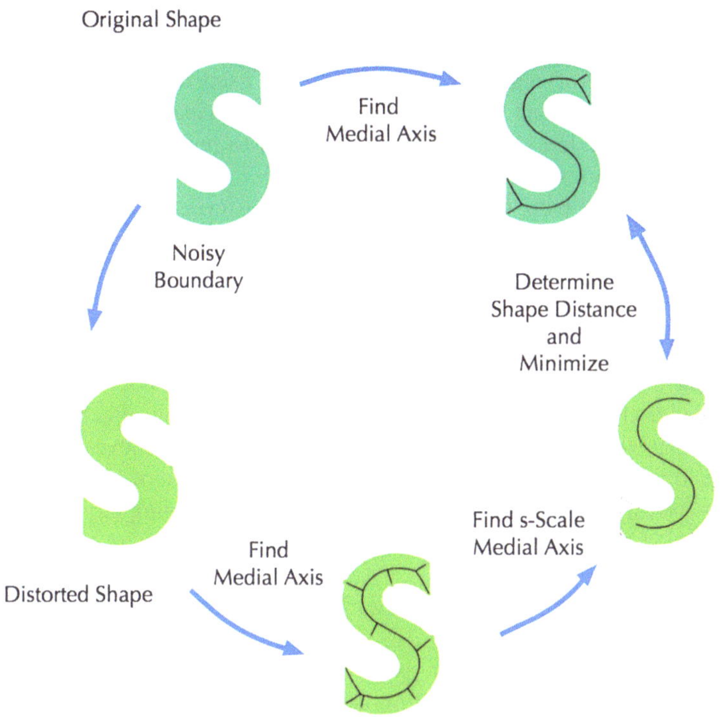

Fig. 8.1 Finding Optimal SAT scale factors: This figure summarizes the methods of this paper. The input is a smooth shape, denoted here as the Original Shape. The medial axis of the Original Shape is then computed and noise is added to the shape boundary. This results in the Distorted Shape. Then, the medial axis of the Distorted Shape is computed, and the s-Scale medial axis is computed for a variety of scales s. The optimal scale is computed by comparing each SAT filtered shape with the Original Shape

8.2.1 The Scale Axis Transform

In order to prune the medial axis of branches not contributing to the defining features of a shape, Bálint Miklós developed the method known as the Scale Axis Transform [9, 10, 17, 18]. The algorithm was inspired by the Medial Axis Transform and proved to be useful in determining the skeleton of a noisy shape.

The medial axis of a shape $\mathcal{O}$, denoted by $\mathrm{M}(\mathcal{O})$, is the set of points x such that there exists a radius r where at least two points on the inscribed ball $B(x, r)$ lie on the shape boundary, $\partial\mathcal{O}$. These balls are known as medial balls; and the medial axis transform of $\mathcal{O}$ is the set of maximal medial balls, where no medial ball provides a cover for another medial ball. That is,

$$\mathrm{MAT}(\mathcal{O}) = \{B(x, d(x, \partial\mathcal{O})) \mid x \in \mathrm{M}(\mathcal{O})\}.$$

Because the medial axis transform is sensitive to the addition of noise to the boundary of the shape $\mathcal{O}$, shapes perturbed by noise will contain branches that are unnecessary for classifying the overall shape. Thus, using the medial axis transform to determine the important and defining features of a shape perturbed by noise can be difficult. The scale axis transform can be used to remove the branches of the medial axis that correspond to noise [9, 10, 17, 18].

Using the medial balls from the medial axis transform, the radii from the balls are scaled by a factor $s > 1$, resulting in the set of balls

$$\mathcal{O}_s = \bigcup_{B(x,r)\in \mathrm{MAT}(\mathcal{O})} B(x, sr).$$

By scaling the medial balls of $\mathrm{MAT}(\mathcal{O})$, smaller balls are covered and are thus disregarded from the medial axis transform of the new shape $\mathcal{O}_s$. Because the centers of the medial balls remain the same, the medial axis of $\mathcal{O}_s$ removes the points that correspond to the covered balls. The scaling s is used to determine the parts of the skeleton corresponding to noise along the boundary. By scaling the medial balls of $\mathrm{MAT}(\mathcal{O}_s)$ by $1/s$ a simplified skeleton is produced. Therefore,

$$\mathrm{SAT}_s(\mathcal{O}) = \{B(x, r/s) \mid B(x, r) \in \mathrm{MAT}(\mathcal{O}_s)\}$$

defines the Scale Axis Transform as the union of the scaled medial balls from $\mathrm{MAT}(\mathcal{O}_s)$ [10, 17, 18]. The algorithm is demonstrated in Fig. 8.2.

8.2.2 *Shape Distance*

Computing the distance between geometric objects has received much attention in recent years, foremost for its utility in computational biology and computational neuroscience. Measuring the differences in the shapes that comprise the human brain, for instance, can offer insight into how diseases affect these regions and can lead to advances in disease detection [11]. In the analysis of the lungs, a new notion of geodesic distance was introduced to be able to compute statistics on the tree structure of airway paths [7, 8]. In this paper, we require a notion of distance that allows us to compare pairs of 2D shapes. This has been approached in previous work through graph edits of the medial axis [21, 24], and by finding diffeomorphisms that map one shape to another in an energy-minimizing manner [4].

To determine a relationship between the level of noise distorting a shape and its optimal scaling under the SAT algorithm, the difference between an original shape and its distorted version must be quantified. Although multiple ways of approaching this task exist, the Fréchet distance between two curves will be used here to numerically evaluate this difference. The Fréchet distance between two curves is frequently referred to as the *dog-walking distance*, because it measures the longest leash required if you were to walk along one curve while your dog walks

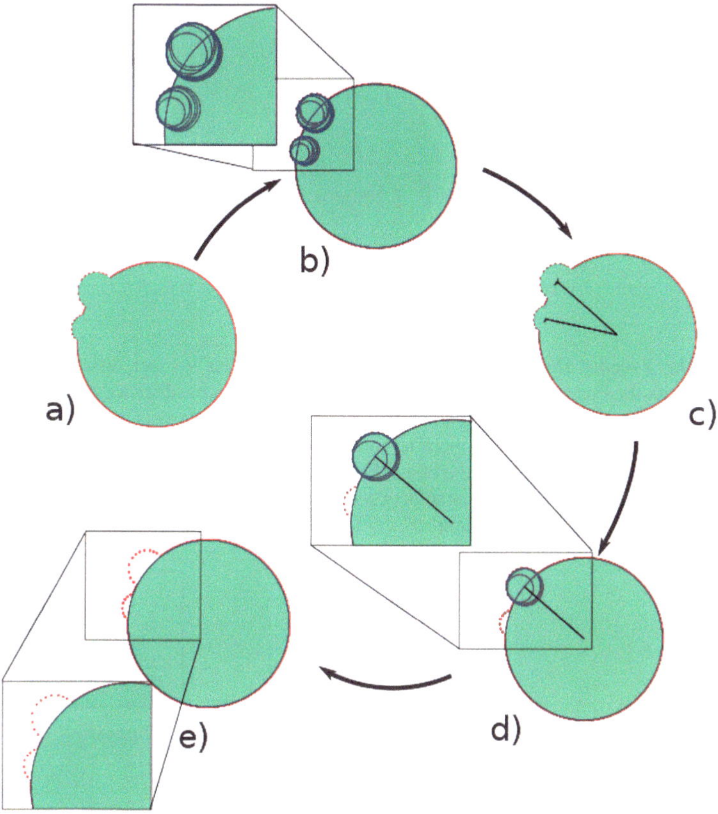

Fig. 8.2 The Scale Axis Transform: This figure demonstrates the Scale Axis Transform algorithm. (**a**) The input or original shape demonstrates exaggerated bumps along the boundary. (**b**) With the (*red*) points along the boundary, the medial balls (denoted by the *blue circles*) are computed. Note that the *green region* is the resulting shape when the medial balls are filled. (**c**) The medial axis (denoted by the *black lines*) is extracted from the centers of the medial balls. With the Scale Axis Transform, spurious branches of the medial axis transform are pruned. Here we demonstrate the effect of the choice for the s scaling parameter of the Scale Axis Transform. (**d**) With $s = 1.2$, the medial balls associated with the smaller bump on the left were filtered, resulting in a simplified shape. (**e**) If s is increased to 1.3, the medial balls associated to both of the bumps are filtered and the shape is simplified to a *circle*

along another path (Fig. 8.3). More concretely, the Fréchet distance between two curves f and g is the infimum over all parameterizations α and β of $[0, 1]$ of the maximum Euclidean distance between $f(\alpha(t))$ and $g(\beta(t))$ for $t \in [0, 1]$ [1] . The advantage of the Fréchet distance is that it is straightforward to compute [1], and works well on the boundaries of 2D simply-connected shapes.

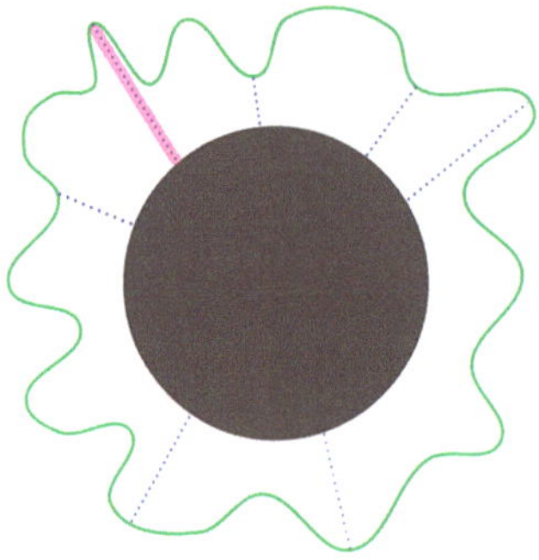

Fig. 8.3 The Fréchet Distance: This figure demonstrates the Fréchet distance between two shapes. As with the Fréchet distance between two curves, the Fréchet distance between two shapes can be thought of as the dog walking distance. Therefore the Fréchet distance between two shapes is the longest leash required if one were to walk along the boundary of the shape and their dog were to walk along the boundary of the other shape. Some candidates for the leash length are given by the *dotted lines*. The Fréchet distance between the boundaries of the *green* and *black shapes* is given by the length of the *pink* segment

Fig. 8.4 This shows the complicated medial axis that results when using an image whose boundary was obtained from a pixelated image. (**a**) Bitmap image. (**b**) Resulting medial axis

8.2.3 Boundary Noise

In this paper, we assume that the noise arises naturally after segmenting the shape boundary from a pixelated 2D image. This type of noise arises as a result of the Mumford-Shah image segmentation process, for example [20]. Moreover, because of the step-like noise added to the shape boundary, the medial axis is complicated by many additional branches that reconstruct these unwanted features. Figure 8.4 demonstrates the type of medial axis that is typically obtained by using shapes with this type of noisy boundary.

Now because image data is commonly used as the basis for extracting geometric shapes, as in medical resonance imaging (MRI), noise from pixelation of the shape boundary is used here. More generally, previous work on filtering the medial axis has considered classes of noise that can include this pixelation error as a special case. For example, in [16], a noisy r-sample is considered, which is a reasonable model for laser range data.

8.2.4 Mesecina

To determine the optimal scaling for the Scale Axis Transform, we relied on a geometric visualization tool developed by Bálint Miklós called Mesecina [10, 17, 18]. Mesecina is an interactive tool that investigates geometric structures in both 2D and 3D. It allows a user to visualize multiple geometric algorithms simultaneously, such as the Power Crust [2], various Delaunay and Voronoi diagrams, and the Scale Axis Transform [10]. This tool offers a quick way to visualize relationships between a variety of fundamental geometric structures.

The input to Mesecina is a point-sampling of the shape boundary. Using Mesecina in 2D, a user can input data by simply drawing points in the window or by importing a point-sampling of the boundary. Interactive data manipulation allows a user to immediately visualize how geometric structures are modified when data points are altered. Further, the parameters of the provided geometric algorithms can easily be modified by the user.

8.3 Optimal Scale Factors

In this section, the procedure for computing optimal SAT scale factors is described. This method combines previous work to suggest reasonable scale factors when applying the Scale Axis Transform to problems where manual tuning is impractical. Given a true shape whose piecewise smooth boundary is known, this process can be decomposed into the following four steps: (A) Perturb the shape boundary by noise; (B) Filter the noisy shape using the SAT algorithm; (C) Compute the Fréchet distance between the boundaries of the original shape and the SAT-filtered noisy shape; and (D) Find the scaling s^* which minimizes the Fréchet distance. The details of each step are explored further in the following paragraphs.

8.3.1 Perturb Shapes by Noise

Let $\mathcal{O}$ denote the original shape, which has a piecewise smooth boundary $\partial\mathcal{O}$. This shape is given as an input to our problem. First, the boundary of this shape is perturbed with pixelation noise. We denote by $\hat{\mathcal{O}}$ the pixelated shape obtained from this rasterizaiton.

Now the level of noise in terms of the Fréchet distance to $\mathcal{O}$, the original shape, is characterized. Let $f = \partial\mathcal{O}$, $g = \partial\hat{\mathcal{O}}$, and let α, β be any parametrizations of $[0, 1]$. Then the **Fréchet distance** between $\mathcal{O}$ and $\hat{\mathcal{O}}$ is given by:

$$\mathrm{F}\left(\mathcal{O}, \hat{\mathcal{O}}\right) = \inf_{\alpha,\beta} \max_{t\in[0,1]} \mathrm{d}\big(f(\alpha(t)), g(\beta(t))\big), \tag{8.1}$$

where $d(x, y)$ denotes the Euclidean distance between points $x, y \subset \mathbb{R}^2$. We abbreviate the Fréchet distance by $\delta_1 = \mathrm{F}(\mathcal{O}, \hat{\mathcal{O}})$. Here, the subscript on δ indicates that the pixelated shape has not yet undergone the SAT filtering (the scale $s = 1$ results in no change).

Recall that perturbing $\mathcal{O}$ by pixelation error results in a noisy medial axis (see Fig. 8.4). We define any circumcenter of the noisy medial axis to be **spurious** if it does not exist in the medial axis of the original un-pixelated shape. For smooth shapes, the medial axis will still have many branches. For example in Fig. 8.1, the Original Shape is the letter S. The medial axis of this shape has four branches, that are used to reconstruct the corners on either end. Our aim in this study is not to compute the simplest medial axis for a given shape but to prune branches corresponding to the noise added.

8.3.2 Filter Noisy Shapes with the SAT Algorithm

Once the noise-perturbed shape $\hat{\mathcal{O}}$ is obtained, we use the Scale Axis Transform to filter the medial axis. For a scale $s > 1$, the SAT-filtered noisy shape is given by

$$\mathrm{SAT}_s(\hat{\mathcal{O}}).$$

This leads to a simplified medial axis, where redundant medial balls have been filtered, as well as a simplified shape resulting from the union of these medial balls. Higher values of s will result in a shape that has been oversimplified, whereas values of s that are too close to 1 will not filter all spurious medial balls.

8.3.3 Compute the Distance to the Original Shape

The distance to the original shape is computed using the Fréchet distance (8.1). The Fréchet distance between the boundaries of the original shape and the SAT-filtered noisy shape is given by:

$$\delta_s = \mathrm{F}\big(\mathcal{O}, \mathrm{SAT}_s(\hat{\mathcal{O}})\big).$$

In applications, one would have access to a set of shapes whose boundaries are perturbed by noise, but not the original shapes. In this paper, we give evidence for which scale factor of the SAT algorithm best filters the medial axis of the original shape when the true shape is known. This information can then be used in cases where the true shape is not known. To determine which scale factor is optimal, the Fréchet distance is used to quantify the distance from the original shape for each SAT scaling. A known feature of the SAT algorithm is that for certain scales $s > 1$, the medial axis may become disconnected. In such cases the Fréchet distance is no longer appropriate, and the Hausdorff distance will be used. The optimal scale will typically occur before any topological changes in the medial axis.

8.3.4 Find the Optimal Scaling

After the previous steps are completed, all that remains is to compute the optimal scale factor to prune the medial axis using the Scale Axis Transform. This is found as

$$\begin{aligned} s^* &= \mathrm{argmin}_{s>1}\, \mathrm{F}\big(\mathcal{O},\ \mathrm{SAT}_s(\hat{\mathcal{O}})\big) \\ &= \mathrm{argmin}_{s>1}\, \delta_s. \end{aligned} \tag{8.2}$$

For each shape $\mathcal{O}$, the noisy shape $\hat{\mathcal{O}}$ is processed with the SAT algorithm using 30 scales $s \geq 1$ to find the minimizing factor s^*. In this study we chose 30 equally-spaced values of s in the range $[1, 2]$ based on the pruning results depicted in [9, 10, 19]. This process is repeated for different image sizes for each shape to establish how the pixel noise affects the optimal scaling for each shape. See Fig. 8.1 for a graphical summary of this process. Note that, in general, we do not expect to find one global optimizer. For non-convex shapes it is possible that the distance described in (C) has more than one local minimizer. We are concerned here with the behavior of the global minimizer across shapes and noise levels. In the next section we present the results of this optimization.

8.4 Results for Six Test Shapes

In our analysis, six shapes were observed: a circle, an ellipse, a cloud, a clover, a hand, and an elephant. All six objects were contained in the unit square to aid in the comparability of the Fréchet distances. In Figs. 8.5–8.10, we present the Fréchet distance between the boundaries of the original shape $\mathcal{O}$ and the SAT filtered noisy shape $\hat{\mathcal{O}}$. Each shape was pixelated with two criteria; their Fréchet distances are depicted in Figs. 8.5–8.10a with the colors black and blue. Notice that $s = 1$ shows the distance between the original shape and the noisy shape before the SAT filtering. This characterizes the level of noise added to each shape.

For simple convex shapes, such as the circle and ellipse (Figs. 8.5 and 8.6), favorable results are produced for a wide range of scalings, typically $s > 1.05$. The Fréchet distance for the circle remains essentially constant for larger scalings. The noise along the boundary of the circle results in a perturbation of the medial center point. For the ellipse, the distance increases with the scaling (Fig. 8.6). This is because for larger scalings, the SAT algorithm will eventually simplify the ellipse

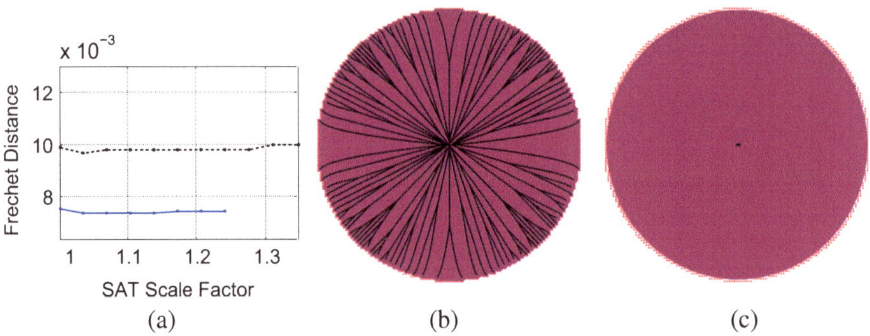

Fig. 8.5 Pictured here is (**a**) the Fréchet distances of the original *circle* and the distorted circle with respect to the scale axis parameter, (**b**) an example of the medial axis of a noisy circle and (**c**) its Scale Axis Transform

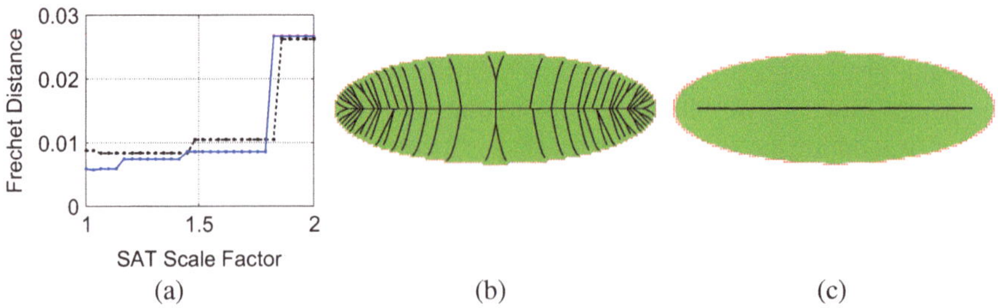

Fig. 8.6 Pictured here is (**a**) the Fréchet distances of the original *ellipse* shape and the distorted ellipse with respect to the scale axis parameter, (**b**) an example of the medial axis of a noisy ellipse and (**c**) its Scale Axis Transform

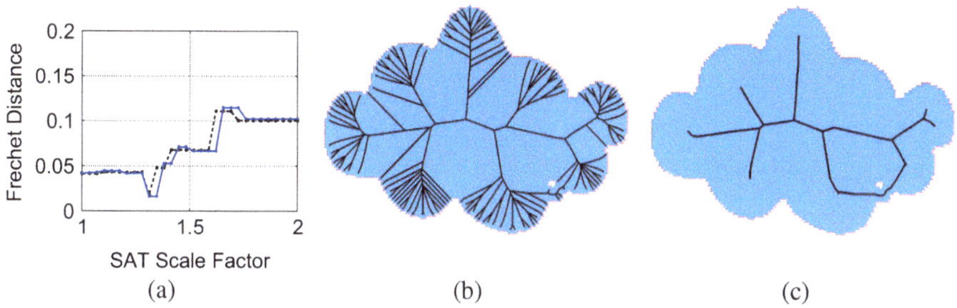

Fig. 8.7 Pictured here is (**a**) the Fréchet distances of the original *cloud* shape and the distorted cloud with respect to the scale axis parameter, (**b**) an example of the medial axis of a noisy cloud and (**c**) its Scale Axis Transform

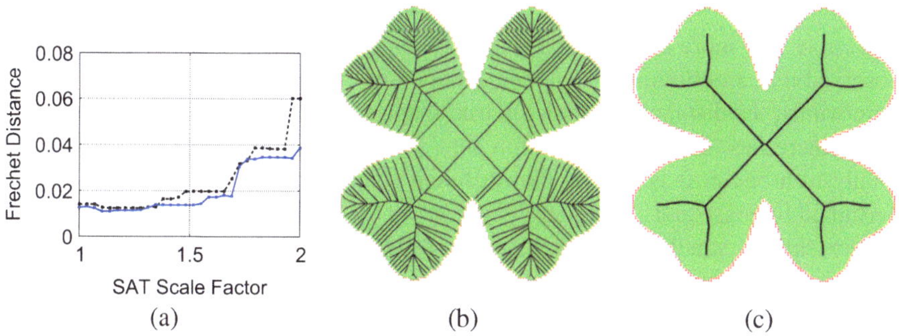

Fig. 8.8 Pictured here is (**a**) the Fréchet distances of the original *clover* shape and the distorted clover with respect to the scale axis parameter, (**b**) an example of the medial axis of a noisy clover and (**c**) its Scale Axis Transform

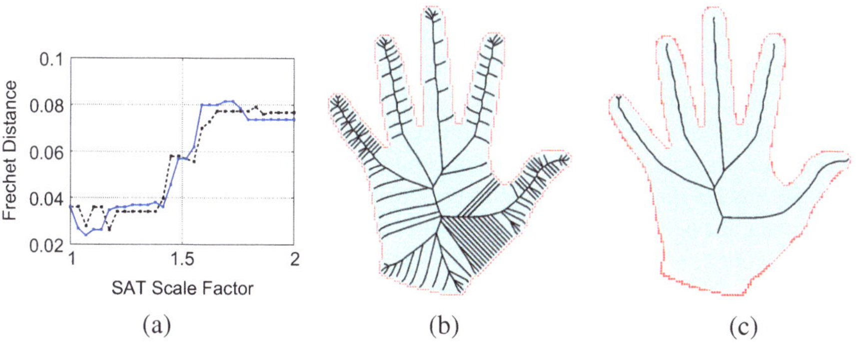

Fig. 8.9 Pictured here is (**a**) the Fréchet distances of the original *hand* shape and the distorted hand with respect to the scale axis parameter, (**b**) an example of the medial axis of a noisy hand and (**c**) its Scale Axis Transform

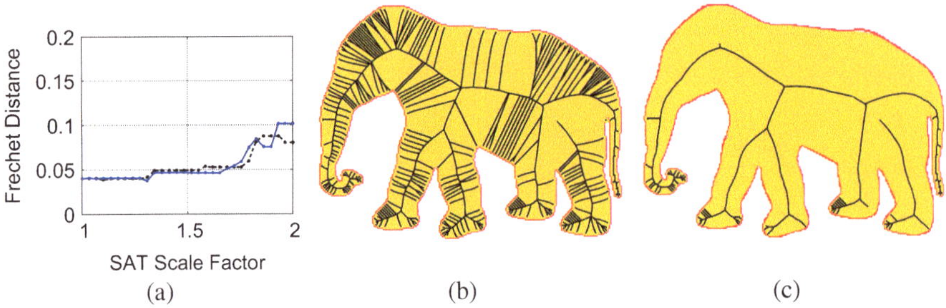

Fig. 8.10 Pictured here is (**a**) the Fréchet distances of the original *elephant* shape and the distorted elephant with respect to the scale axis parameter, (**b**) an example of the medial axis of a noisy elephant and (**c**) its Scale Axis Transform

to a circle. For more complicated shapes (Figs. 8.7–8.10), such as the cloud, it was observed that a minimum distance is obtained in the range of $s = 1.1$ to $s = 1.3$. Afterwards, the distance between the true shape and the reconstruction of the noisy shape increases. The SAT algorithm first simplifies the cloud to an ellipse, and finally to a circle. It is natural that different shapes will behave differently when perturbed by noise and subsequently filtered by the SAT algorithm. For highly non-convex shapes, smaller scalings (e.g., closer to $s = 1.1$ than $s = 1.3$) should be favored to avoid topological changes in the medial axis.

8.5 Discussion

In this paper, we investigated optimal scaling factors for the Scale Axis Transform on a set of six shapes. The results gave insight into how well the medial axis of a noisy shape can be filtered of its spurious branches using different scale factors for the SAT algorithm. As a rule of thumb, $s \in [1.1, 1.3]$ produced favorable results and is well-suited for use in cases where tuning s to each shape is impractical.

Further, we believe that our results can be expanded. Although the focus here was the Fréchet distance, other notions of shape distance can be explored. As was mentioned, the Scale Axis Transform prunes the spurious branches from the medial axis. Therefore, a natural next step is to explore the similarity of the two shapes by comparing their skeletons [21, 22]. In addition, shape distance can also be defined with a diffeomorphism mapping from one shape to another [4]. Therefore, other shape distances can be used to further investigate the optimal scaling of the Scale Axis Transform.

Although only 2D shapes were explored here, a study of 3D shapes is another step that can be examined. Since the Scale Axis Transform also applies to 3D shapes, we believe our methodology, with the use of the Fréchet distance for 3D shapes [5], can be used to study the optimal scale factors for 3D images. As with our 2D study, this extension will be useful in filtering images susceptible to pixelation noise, as is often seen in 3D medical scans. Thus, a similar study in 3D may be valuable.

References

1. Alt, H., Godau, M.: Computing the Fréchet distance between two polygonal curves. Int. J. Comput. Geom. Appl. **5**, 75–91 (1995)
2. Amenta, N., Choi, S., Kolluri, R.K.: The power crust. In: Proceedings of the Sixth ACM Symposium on Solid Modeling and Applications, Ann Arbor pp. 249–266. ACM (2001)
3. Attali, D., Montanvert, A.: Computing and simplifying 2D and 3D continuous skeletons. Comput. Vis. Image Underst. **67**(3), 261–273 (1997)
4. Beg, M.F., Miller, M.I., Trouvé, A., Younes, L.: Computing large deformation metric mappings via geodesic flows of diffeomorphisms. Int. J. Comput. Vis. **61**(2), 139–157 (2005)
5. Buchin, K., Buchin, M., Wenk, C.: Computing the Fréchet distance between simple polygons in polynomial time. In: Proceedings of the Twenty-Second Annual Symposium on Computational Geometry (SCG '06), Sedona, pp. 80–87. ACM (2006)
6. Chazal, F., Lieutier, A.: The λ-medial axis. Graph. Models **67**(4), 304–331 (2005)
7. Feragen, A., Lauze, F., Lo, P., de Bruijne, M., Nielsen, M.: Geometries on spaces of treelike shapes. In: Computer Vision–ACCV 2010, Queenstown, pp. 160–173. Springer (2011)
8. Feragen, A., Owen, M., Petersen, J., Wille, M.M., Thomsen, L.H., Dirksen, A., de Bruijne, M.: Tree-space statistics and approximations for large-scale analysis of anatomical trees. In: Information Processing in Medical Imaging, Asilomar, pp. 74–85. Springer (2013)
9. Giesen, J., Miklos, B., Pauly, M., Wormser, C.: The scale axis picture show. In: ACM Video/Multimedia Session of Symposium of Computational Geometry (2009)
10. Giesen, J., Miklos, B., Pauly, M., Wormser, C.: The scale axis transform. In: Proceedings of the 25th Annual Symposium on Computational Geometry, Aarhus, pp. 106–115. ACM (2009)
11. Klein, A., Andersson, J., Ardekani, B.A., Ashburner, J., Avants, B., Chiang, M.C., Christensen, G.E., Collins, D.L., Gee, J., Hellier, P., et al.: Evaluation of 14 nonlinear deformation algorithms applied to human brain MRI registration. Neuroimage **46**(3), 786–802 (2009)
12. Leonard, K.: An efficiency criterion for 2D shape model selection. In: IEEE Computer Society Conference on Computer Vision and Pattern Recognition, New York, vol. 1, pp. 1289–1296. IEEE (2006)
13. Leonard, K.: Efficient shape modeling: entropy, adaptive coding, and boundary curves-vs-Blum's medial axis. Int. J. Comput. Vis. **74**(2), 183–199 (2007)
14. Lieutier, A.: Any open bounded subset of $\mathbb{R}^n$ has the same homotopy type as its medial axis. Comput. Aided Design **36**(11), 1029–1046 (2004)
15. Liu, L., Chambers, E.W., Letscher, D., Ju, T.: Extended grassfire transform on medial axes of 2D shapes. Comput. Aided Design **43**(11), 1496–1505 (2011)
16. Mederos, B., Amenta, N., Velho, L., de Figueiredo, L.H.: Surface reconstruction for noisy point clouds. In: Symposium on Geometry Processing, Vienna, pp. 53–62. Citeseer (2005)
17. Miklós, B.: The scale axis transform. Ph.D. thesis, ETH Zürich (2010)
18. Miklós, B., Giesen, J., Pauly, M.: Medial axis approximation from inner Voronoi balls: a demo of the Mesecina tool. In: Proceedings of the 23rd Annual Symposium on Computational Geometry, Gyeongju. ACM (2007)
19. Miklos, B., Giesen, J., Pauly, M.: Discrete scale axis representations for 3D geometry. ACM Trans. Graph. (TOG) **29**(4), 101 (2010)
20. Mumford, D., Shah, J.: Optimal approximations by piecewise smooth functions and associated variational problems. Commun. Pure Appl. Math. **42**, 577–685 (1989)
21. Sebastian, T., Klein, P., Kimia, B.: Recognition of shapes by editing shock graphs. In: IEEE International Conference on Computer Vision, Vancouver, vol. 1, pp. 755–755. IEEE Computer Society (2001)
22. Sebastian, T.B., Klein, P.N., Kimia, B.B.: Recognition of shapes by editing their shock graphs. IEEE Trans. Pattern Anal. Mach. Intell. **26**(5), 550–571 (2004)
23. Shaked, D., Bruckstein, A.M.: Pruning medial axes. Comput. Vis. Image Underst. **69**(2), 156–169 (1998)
24. Siddiqi, K., Shokoufandeh, A., Dickinson, S.J., Zucker, S.W.: Shock graphs and shape matching. Int. J. Comput. Vis. **35**(1), 13–32 (1999)

Chapter 9
Identifying Perceptually Salient Features on 2D Shapes

Lisa J. Larsson, Géraldine Morin, Antoine Begault, Raphaëlle Chaine, Jeannine Abiva, Evelyne Hubert, Monica Hurdal, Mao Li, Beatriz Paniagua, Giang Tran, and Marie-Paule Cani

L.J. Larsson (✉)
Department of Mathematics and Statistics, McGill University, 805 Sherbrooke Street West, Montreal, QC H3A 0B9, Canada
e-mail: lisa.powers@mail.mcgill.ca

G. Morin
University of Toulouse - INPT - IRIT, IRIT-N7, 2 Rue Camichel, 31071 Toulouse, France
e-mail: geraldine.morin@enseeiht.fr

A. Begault • M.-P. Cani
LJK, University Grenoble Alpes, CNRS and Inria, F-38000 Grenoble, France
e-mail: antoine.begault@inria.fr ; marie-paule.cani@inria.fr

R. Chaine
Université de Lyon, Université Lyon 1, CNRS, LIRIS, UMR5205, F-69622 Lyon, France
e-mail: raphaelle.chaine@liris.cnrs.fr

J. Abiva
Department of Mathematics, University of Iowa, 14 MacLean Hall, Iowa City, IA 52242, USA
e-mail: jeannine.abiva@gmail.com

E. Hubert
INRIA Méditerranée, 2004 Route des Lucioles, Sophia Antipolis 06902, France
e-mail: evelyne.hubert@inria.fr

M. Hurdal • M. Li
Department of Mathematics, Florida State University, 208 Love Building, 1017 Academic Way, Tallahassee, FL 32306, USA
e-mail: mhurdal@math.fsu.edu; mli@math.fsu.edu

B. Paniagua
Department of Psychiatry, Computer Science and Orthodontics, University of North Carolina at Chapel Hill, Medical School Wing C, Third Floor, Suite 356 Campus Box CB 7160, Chapel Hill, NC 27599, USA
e-mail: beatriz_paniagua@med.unc.edu

G. Tran
Department of Mathematics, University of California, Los Angeles, 520 Portola Plaza, Los Angeles, CA 90095, USA
e-mail: giangtran@math.ucla.edu

K. Leonard, S. Tari (eds.), *Research in Shape Modeling*, Association for Women in Mathematics Series 1, DOI 10.1007/978-3-319-16348-2_9

Abstract Maintaining the local style and scale of 2D shape features during deformation, such as when elongating, compressing, or bending a shape, is essential for interactive shape editing. To achieve this, a necessary first step is to develop a robust classification method able to detect salient shape features, if possible in a hierarchical manner. Our aim is to overcome the limitations of existing techniques, which are not always able to detect what a user immediately identifies as a shape feature. Therefore, we first conduct a user study enabling us to learn how shape features are perceived. We then propose and compare several algorithms, all based on the medial axis transform or similar skeletal representations, to identify relevant shape features from this perceptual viewpoint. We discuss the results of each algorithm and compare them with those of the user study, leading to a practical solution for computing hierarchies of salient features on 2D shapes.

9.1 Introduction

Natural objects typically include a hierarchy of shape features, from fine-scale details to the main object. These features, such as those in the examples of Fig. 9.1, are fundamental to our perception of the object. Similarly, digital models of shape incorporate a number of features of various scales. During interactive shape editing, artists should preferably not have to manually adjust or duplicate all these features when deforming the parent shape, e.g., by bending, twisting, stretching or shrinking it: this would be time-consuming, even for skilled artists. Therefore, research in interactive shape design has sought methods that enable adjustments on the parent shape while automatically preserving the style and aesthetics of small shape features.

This paper takes a necessary step towards this challenging goal: it proposes perceptually-based algorithms to detect shape features, in the case of 2D shapes. Once shape features are detected, they can be carefully manipulated during deformation processes. In this work, our first focus is defining what is meant by a shape feature. Therefore, we use the results of a perceptual user study to guide our definition of shape features. Secondly, we choose to use the medial axis transform to detect features on 2D shapes, since this is likely to ease subsequent generalization to 3D shapes, compared to other criteria, such as local curvature.

Fig. 9.1 Real shapes with perceptually salient features, such as seeds on strawberries, ridges on leaves, and skin details on a chameleon

Throughout this paper, $\mathcal{S}$ will denote a shape in $\mathbb{R}^2$ that has a closed, piecewise smooth boundary. As a subset of $\mathbb{R}^2$, $\mathcal{S}$ is a compact set.

The organization of this paper is as follows: In Sect. 9.2, we present previous work. A user study conducted to understand human perception of shape features is then motivated and described in Sect. 9.3; the results of this study are also presented. Then, two classes of algorithms for detecting shape features are proposed: a geometric algorithm is described in Sect. 9.4, followed by several methods enabling the detection of features at different resolutions in Sect. 9.5. Section 9.6 then presents numerical tests of the new algorithms on a variety of interesting shapes and compares this to an existing feature-detection algorithm. We conclude with a discussion of these results and directions for future work.

9.2 Previous Work

9.2.1 The Medial Axis Transform

The medial axis can be defined equivalently in different ways, for instance, in terms of maximally inscribed balls or the shock set of the eikonal flow from the shape boundary [4, 7, 8, 16]. The following definition was presented in [3, 4].

Definition 9.1 (Medial Axis Transform). The Medial Axis Transform of $\mathcal{S}$ is given by the set of locations $\mathcal{M}$ internal to the object with more than one corresponding closest boundary point and their distance $\mathcal{R}$ from the boundary $\partial\mathcal{S}$.

Remark. When the shape $\mathcal{S}$ is non-convex, an "exterior medial axis" can also be defined. These are points in $\mathcal{S}^c$ that are equidistant to two or more closest points along $\partial\mathcal{S}$.

The shape $\mathcal{S}$ can be reconstructed as the union of balls centered along $\mathcal{M}$ of radius $\mathcal{R}$. These balls are maximal and have at least two contact points on the surface. The exterior medial axis is not required for shape reconstruction.

It is often convenient to assume that $\mathcal{S}$ is simply connected; i.e., that it has no holes. This simplifies the topology of the medial axis, in particular guaranteeing no loops. In the remainder of this paper, we only focus on such simply connected shapes.

The medial axis representation has been widely studied in computational geometry. In practice, the medial axis can be approximated as a subset of the Voronoi diagram of points sampled on the boundary of the shape [1]. In [13], the authors edit the shock graph to systematically define the distance between two shapes for recognition.

(a) (b)

Fig. 9.2 Shapes with salient features. (**a**) The bumps protruding from the large sphere are salient. (**b**) A shape with many salient features

9.2.2 *Identifying Salient Features*

In his seminal work [9], Michael Leyton characterizes features on 2D shapes using curvature of the shape boundary: a salient feature of a 2D shape is identified by the presence of two extrema of curvature of the same sign along the boundary, $\partial \mathcal{S}$. The two extrema should occur at the contact points of one medial ball of the medial axis (or of the external medial axis).

Figure 9.2 shows a number of salient shape features that follow this definition: the shape in (b) has many salient features; some are larger, such as what appears to be the head of a bird, and others are smaller in scale. In our user study, we probe whether, on average, parts like the bird's head are perceived to be a shape feature or part of the main shape. We do not restrict ourselves to any particular definition of salience here, but rather use the results of the perceptual study to guide our definition of shape feature.

Directly using skeletal representations to detect salient features has been proposed in previous work. For example, in [12], the authors use a modification of the medial axis transform, called the chord axis transform, and decompose shapes based on the extremal chord strength. In [18], the authors use both skeletal and boundary features and define a protrusion strength that is then used to decompose the shape. The originality of our approach compared to these methods is the fact that we build our work on a user study, enabling us to address the identification of perceptually salient shape features.

Furthermore, there are many algorithms designed for pruning the medial axis that can be applied to identifying shape features. These algorithms were developed because the medial axis transform is known to be very sensitive to noise on the shape boundary, resulting in spurious branches [1, 2, 5, 15, 17]. Much work has been devoted to the pruning of these non-informative branches. The resulting algorithms, aimed at detecting features in order to remove the less salient ones, should indeed be considered with respect to our goal. These approaches filter points on the medial axis depending on the geometric configuration of the contact points of the medial balls (e.g., the radius of medial balls combined with the angle between a center and the contact points [2, 15], the radius of the ball circumscribing the contact points [5], or the area enclosed by the contact points [14, 17]). More recently the Scale Axis Transform (SAT) was developed, which uses a nonlinear scaling of the medial balls to produce a hierarchy of simplified medial axes [6, 11]. This method is recalled in more detail in Sect. 9.5, as we evaluated it with respect to the results of our user study, and then build on it to propose a solution that better matches our goals.

To adhere to the results of the user study, we propose to extend the SAT in order to take into account the length or thickness of medial branches. To compute the length, we rely on the extended distance function (EDF) [10], which measures the tubularity of a shape. A detailed description of EDF is given in Sect. 9.5; to compute thickness, WEDF, an original weighted version of EDF, is then proposed. Combining these branch characteristics with the SAT algorithm is one of the key points of our solution.

9.3 User Study on Shape Feature Perception

A user study was conducted to determine the criteria for the perceptual identification of features on a 2D shape. For that, given a collection of 2D shapes, we asked the users to identify any shape features they considered different from the main shape. This study was designed to gauge the consistency with which users identify shape features, and both motivates and validates the algorithms we construct. The user study consisted of a set of 44 shapes, that were printed on three sheets of paper. The results for shapes with semantic content—i.e., shapes that looked like plants, animals, or other naturally-occurring objects—were discarded to mitigate semantic bias. The order and the orientation of the shapes was random, to reduce the influence of similar shapes in adjacent positions.

Indeed, our aim in this study was to identify the geometric criteria that are important in identifying shape features. For the non-specific, simply connected shapes presented in the user study, we assumed that the partition of the shape into a main shape plus shape features was independent from the shape's orientation. Before conducting the user study, we formed four hypotheses about the properties of shape features in terms of the radius and orientation of the shape's medial axis:

(H1) *Any part of a shape delimited by two corresponding extrema of curvature of the same sign (i.e., occurring at the contact points of one medial ball), is a shape feature. This is Leyton's criteria [9].*

(H2) *A junction of the medial axis (local Y shape) is important but neither necessary nor sufficient to characterize a shape feature.*

(H3) *A change in thickness along the medial axis is more predictive of shape features than a change in direction. Yet, it is neither necessary nor sufficient to characterize a shape feature.*

(H4) *If the main shape is not of larger radius than the features, it is of larger length. Moreover, a part with smaller area is more likely to be a shape feature than a part with smaller length.*

Once the users obtained the set of printed shapes, we gave them the following instruction:

(Q) *We are trying to construct a hierarchy of a shape's features. Highlight what you think is NOT the main shape.*

The user study was given to 24 people, the sheets were scanned and aligned, and each pixel was averaged over the 24 forms. The average results in grayscale are presented in Fig. 9.20. A summary of relevant results is given below:

The first hypothesis aims to verify the definition of salient shape feature given by Leyton (see Sect. 9.2). This hypothesis was both validated and refuted by the user study. Actually, Figs. 9.3a, b confirm (H1), while Fig. 9.3c and Fig. 9.3d refute it.

In Fig. 9.3a, nearly all users identify the nob on top as a shape feature, while the white area is the main shape. This feature is delimited by two extrema of curvature of the same sign. Figure 9.3b also confirms (H1). Although some users thought of the lower and upper extremities as shape features, the majority considered those parts to be the main shape, while the only feature delimited by two extrema of curvature of the same sign was chosen to be a shape feature by the majority of the users. In Fig. 9.3c, all four extremities should be considered shape features under (H1). However, this is not observed, as the two extremities that are aligned are considered by the majority of users to be the main shape. In Fig. 9.3d, the users both validate and refute (H1). The small nob on the left side of the shape is overwhelmingly characterized as a shape feature, this is in agreement with (H1). However, both the large white areas are delimited by two extrema of curvature of the same sign, yet they are considered the main shape. This is likely because of the larger size of these areas compared to the thin connecting strip. In summary, there are cases that both support and refute this hypothesis, so it cannot be used in isolation to identify shape features.

The second hypothesis is concerned with junctions in the medial axis. (H2) conjectures that a junction in the medial axis is important but neither sufficient

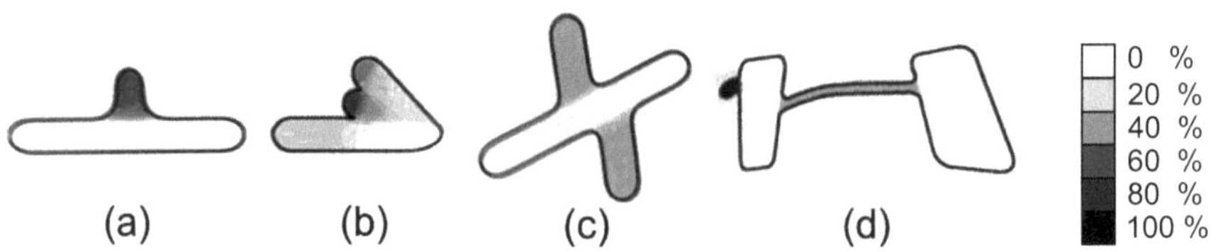

Fig. 9.3 Results of the user study for (H1). (From the *left*) The first two shapes validate (H1), as the majority of users choose shape features that are consistent with Leyton's definition of salience. The third shape disagrees with Leyton's definition and (H1), as only two of the four salient parts are identified as shape features. The fourth shape also disagrees with (H1), as the two larger areas are salient features by this definition

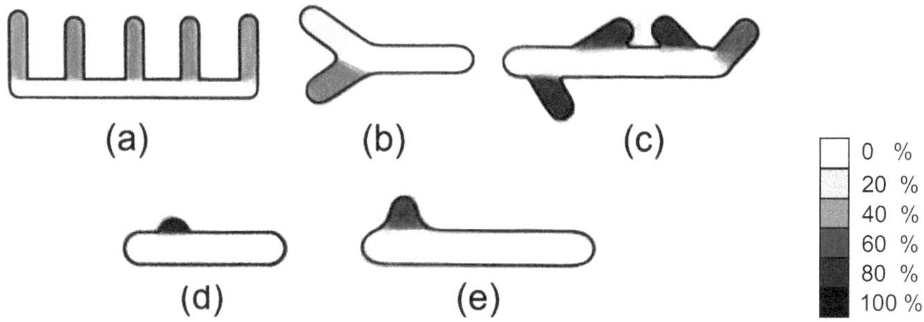

Fig. 9.4 Results of the user study for (H2). (From the *top left*) The first shape (**a**) confirms (H2) because of the first and last prong of the comb—there is no junction here, yet they are shape features. The shape in (**b**) also confirms (H2) because it was unclear which branch of the junction was the shape feature—many thought it was all the main shape. Shape (**c**) confirms (H2), as there is no junction for the right-most feature. In the *bottom* two figures, a junction is sufficient

nor necessary to characterize shape features. A junction refers to a meeting point of three edges of the medial axis. The results here were quite interesting. There were many cases (see Fig. 9.4) where a junction from the primary medial axis branch was sufficient to characterize a shape feature (for example, Fig. 9.4d and Fig. 9.4e. However, for the Y shape (Fig. 9.4b) the users did not reach a consensus about which branch of the Y was the shape feature, and some users found there to be no shape features. It is difficult to draw broad conclusions from this as the Y shape has semantic information—it is shaped like a letter, which may influence the perception of shape features in this case. In the top row, Fig. 9.4a, c have repeated shape features. In the comb example (Fig. 9.4a) all the prongs are shape features, even though the first and last do not occur at a junction of the medial axis. Similarly, in Fig. 9.4c the right-most feature does not occur at a junction of the medial axis, so a junction is not necessary in this case. These results demonstrate that junctions in the medial axis are strong indicators of shape features, but not sufficient, and suggest that repeated features should also be taken into account. Hypothesis (H2) was therefore validated by the user study.

The third hypothesis considers whether it is more important to have a change in shape thickness or a change in the direction of the shape to classify shape features. In Fig. 9.5, there are three examples that support (H3). The results were not unanimous: In Fig. 9.5a, some users highlighted the bottom left part as a feature, but most decided that the main shape was the V to the left, and the feature was the nob to the right. The left V has a consistent thickness, but not a consistent orientation, while the thickness of the right nob is significantly reduced. This supports (H3). Figure 9.5b also depicts mixed opinions, though the majority of users selected the thin right part of the shape to be a shape feature, which is the part with the smallest thickness. The results were similar for Fig. 9.5c. These results validate (H3). The example of the

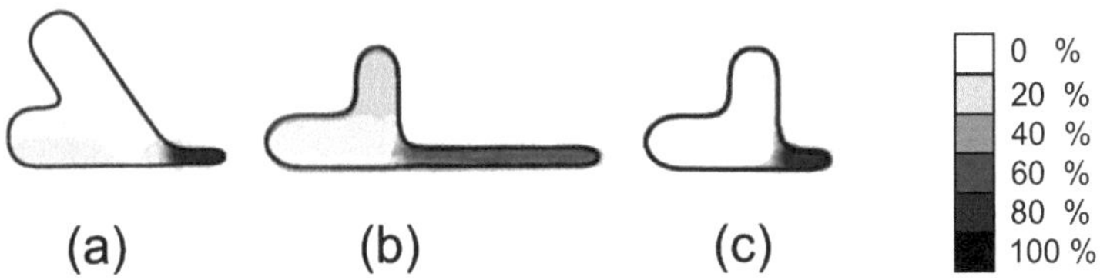

Fig. 9.5 Results of the user study for (H3). All three figures confirm (H3), which says that a change in thickness is more important than a change in direction in identifying shape features

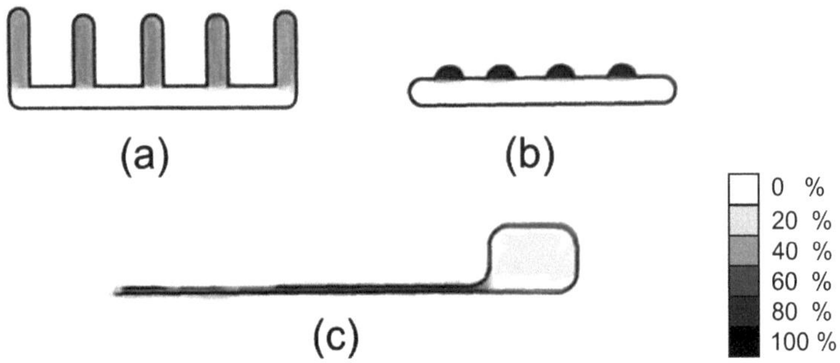

Fig. 9.6 Results of the user study for (H4). All three figures confirm (H4), which says that the measure (in length or area) of the main shape is larger than that of the shape feature

comb shape (Fig. 9.4a) illustrates that changes in thickness are not necessary nor sufficient to determine a shape feature. The thickness of the main part is similar to the thickness of the features in this example.

The fourth hypothesis addresses the case where the main shape is not as thick or as long as the shape features. Figs. 9.6a, b show cases where the main shape has the same thickness as the shape features. This validates (H4), as in these two cases, the users identified shape features that have the smallest length. Therefore, given a shape with nearly constant thickness, it is necessary and sufficient to use length to successfully identify shape features. In Fig. 9.6c, a shape with non-constant thickness is shown. In this case, length alone is not sufficient to characterize shape features, as the thin feature is much longer than the main shape. For shapes with non-constant thickness, a branch with a bigger area is more likely to be considered the main shape.

In summary, these results show that:

- Two extrema of curvature of the same sign do not always identify shape features.
- Junctions in the medial axis are important for determining shape features, but are neither necessary nor sufficient.
- Repeated parts can be considered features even when there is no junction in the medial axis.
- Change in radius is more important than change in orientation to identify shape features.

- The length of a part of a shape is necessary and sufficient to identify shape features when the thickness of the shape is nearly constant.
- A part of a shape with larger area is more likely to be considered the main shape than a lengthier part of smaller area.

These conclusions have driven the construction of a geometric feature-detection algorithm, based on the radius and orientation of the medial axis, which is presented next.

9.4 Perceptually-Based Geometric Feature Detection

Our goal is to partition a 2D shape to separate the main part of the shape from the shape features. The geometric approach presented in this section directly builds on the results of the user study we just described: the input of the algorithm is a 2D shape and the output is the features, given in terms of their medial axis representation. The method involves first computing the medial axis, and then partitioning it using criteria that combine branching information with variations in radius of the medial balls.

9.4.1 Junctions in the Medial Axis

The most natural way to partition a 2D shape according to its medial axis is to consider points where the topology of the medial axis changes, that is, to categorize the branching parts as features. In Fig. 9.7, the small branch of the medial axis corresponds to the raised bump on the main shape.

However, as confirmed by (H2) in the user-study, determining which branch corresponds to the feature and which branch corresponds to the main shape requires extra information in addition to the detection of junctions: in Fig. 9.7, there are three branches of the medial axis, and while two correspond to what is visually perceived as the main shape, one corresponds to the shape feature. Considering junctions of the medial axis in isolation is therefore not sufficient. In practice, we combine it with other information, namely changes of radius and of orientation among neighboring branches. This corresponds to the conditions for identifying features in Algorithm 1 below.

Fig. 9.7 The medial axis (in *white*) for a shape (in *blue*). The junction in the medial axis indicates a branching, which is a criteria for identifying shape features

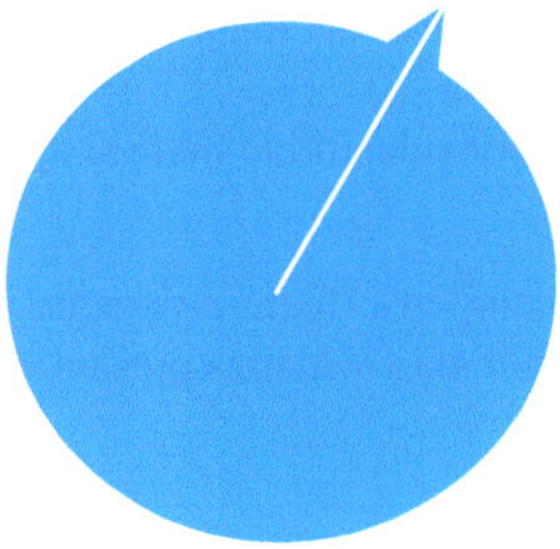

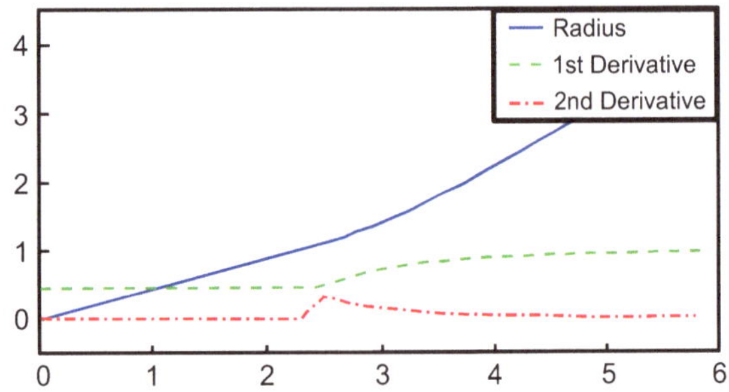

Fig. 9.8 *Left*: Although the medial axis has no junction, the nose is a shape feature. *Right*: The radius and its derivatives are shown. The calculation starts at the tip of the feature and proceeds towards the circle center

9.4.2 Changes in Radius

As demonstrated in the user study (H2), considering junctions is important but neither sufficient nor necessary. Shape features may also arise along a single branch of the medial axis. Figure 9.8 depicts a simple example, namely a sphere with a small triangular outcropping, which can appear as a tail or a nose. The nose is a shape feature, however there is no branch in the medial axis that would lead to its detection. This suggests that along each branch, the radius must be considered, and the branch will be further subdivided if there is a significant change in thickness. A subdivision of the branch means that we split the branch into two branches and add a new junction between the two.

One way to analyze the thickness is to consider the variation of the radius on the medial axis. Analyzing the behavior of the medial axis in Fig. 9.8 one can observe that there is a sudden increase in the second derivative (see the curves in Fig. 9.8, right). Thus, we propose to use the variation of the radius along the curve to further partition the medial axis. New junctions are added on simple branches when a second order variation of the radius is detected.

9.4.3 Resulting Geometric Algorithm

Once the medial axis has been partitioned by considering both junction and thickness information along branches, consistency of thickness and orientation at each junction are considered sequentially to compute the shape decomposition we are looking for. According to (H3), we first consider changes in thickness, and when there is no change in thickness, we identify features based on directions. Our method is summarized in Algorithm 1 below. Note that we implemented a discrete version of the radius-based partitioning method just discussed: the medial

Algorithm 1 Geometric feature detection algorithm

Input: Given $\mathcal{S}$, and a three thresholds ε, T and α used for the second order variation of the radius of branches, the relative change of thickness, and the relative angle between branches at a junction, respectively.

Compute $\mathcal{M}$, the medial axis.

Preprocessing:
if there is a second order change of radius along the skeleton, that is $\Delta_2 R = R_{i+1} - 2R_i + R_{i-1} > \varepsilon$ **then**
  Split the branch at this point (add a junction).
end if

Main Loop:
for each junction x in the medial axis **do**

  Part I: Compare the radius between adjacent branches
  for each branch b_i entering the junction x **do**
    1. Get the intersection point x_i of b_i with the circle centered at x of radius $R(x)$.
    2. Compute the radius of the ith branch $r_i = R(x_i)$.
  end for

  Find $r = \max_i r_i$ the maximum radius.
  for each branch b_i associated with the junction x **do**
    if $r - r_i > T$ **then**
      Label b_i a feature.
    else
      Keep b_i as the main shape.
    end if
  end for

  Part II: Compare the orientation between adjacent branches
  if All branches (b_i) are labeled as the main shape **then**
    for each branch b_i associated with the junction x **do**
      Compute the orientation θ_i of the branch b_i at x_i.
      if for all $j : |\theta_i - \theta_j| - \pi > \alpha$ **then**
        Label b_i a feature.
      else
        Keep b_i the main shape.
      end if
    end for
  end if
end for

axis is sampled into a graph of nodes with associated radius. Adjacent nodes refer to neighboring points along this discretized graph. The second order derivative of the radius is approximated by the second discrete differences (denoted by Δ_2) on three successive values of the radius along a branch.

Results of this algorithm will be discussed in Sect. 9.6. This algorithm requires different parameter values for comparing changes of radius along a branch (Preprocessing), branch radii at junctions (Part I), and branch orientations (Part II). These parameters can be tuned to get results that consistently match the user study. Moreover the algorithm provides the relative importance of adjacent branches at each junction: this leads to different levels of resolution. Of course, branches of the same level are not necessarily topologically connected. Moreover, branch importance is computed based on local information at each junction. A drawback of this method could be its lack of computing global branch importance: branches of the same level may be very different in thickness, for example. Using this algorithm to detect hierarchies of shape features at different resolutions would require complex tuning, with different families of values for the three parameters. We therefore investigate an approach that uses a single resolution threshold, presented next.

9.5 Towards Multi-resolution Feature Detection

In this section, we develop an algorithm that can identify a hierarchy of shape features of different scales. The starting point to achieve this goal is the Scale Axis Transform, which achieves multiscale feature detection using a nonlinear scaling of the medial balls. We first study the standard SAT algorithm and identify why it does not match our goals. We then propose two extensions, based on different weightings of the scaling function that use length and area criteria, respectively. The input to the problem is again the 2D shape, and the output is a hierarchy of features based on a thresholding parameter.

9.5.1 Using the Scale Axis Transform

The *Scale Axis Transform* is a method for pruning spurious branches from the medial axis of 2D and 3D shapes [6, 11]. Given a shape, $\mathcal{S}$, the first step of the algorithm is to compute its medial axis $\mathcal{M}$. Next, each ball of the medial axis is scaled by a factor $s > 0$ (thus creating a non-linear scaling of $\mathcal{S}$), and the union of these scaled medial balls forms the s-scaled version of the shape $\mathcal{S}$. If $s > 1$, then the shape is dilated, if $s < 1$, then the shape is shrunk; $s = 1$ recovers the original shape. Under a dilation, $s > 1$, there will be medial balls that no longer contribute to the boundary of the s-scaled shape. These medial balls are removed. The inverse scaling $\frac{1}{s}$ is then applied to the new medial balls, recovering a medial axis representation of $\mathcal{S}$ where the parts of the medial axis corresponding to redundant medial balls are now pruned. This method is very effective for pruning parts of the medial axis that are artifacts of rasterized image boundaries and other boundary noise. Additionally, when $s > 1$ is progressively increased, this method yields a hierarchy of medial axes that correspond to progressively simpler versions of the input shape, $\mathcal{S}$.

Fig. 9.9 Shapes with features (bumps) of the same radius as the main shape. The radius of the medial axis transform is nearly constant throughout the shape: the bumps will thus not be detected as salient features using the classical SAT. However, in such cases, the user study clearly indicates that the bump should still be identified as a feature, even without repetition

In [6], the authors show that for any scaling $0 < s < 1$, the s-scaled shape is homotopy equivalent to the input shape $\mathcal{S}$. Moreover, for a class of scalings $s > 1$, the s-scaled shape is also shown to be homotopy equivalent. The s-scaled shape can be obtained as the appropriate level set of a multiplicatively weighted distance function, where the weight is the distance from the surface to the closest point on the medial axis—the radius of the corresponding medial ball. These homotopy results are valid in any spatial dimension. In [11], the authors extend the algorithm for computing pruned medial surfaces in 3D.

Could the Scale Axis Transform be used for perceptual feature detection? Indeed, defining a hierarchy of shape features according to their persistence under the SAT non-linear scaling looks natural. However, since the SAT prunes features based on the relative size of their medial balls as compared to neighboring balls, a shape part will be identified as the *main shape* only if it is of larger radius than neighboring parts. For example, in Fig. 9.9, the main part of the shape is of the same radius as the features, meaning the SAT will fail to identify the features correctly—the whole shape will be considered the main shape. Features that are attached to a support with the same radius will not be pruned using the Scale Axis Transform. The next two sections propose modifications of the SAT aimed at improving detection of shape features when the part perceived as the main part of the shape is relatively thin.

9.5.2 *Length-Weighted SAT*

Results of the perceptual user study (H4) showed that the main shape may not be of larger radius than features, but that it should be associated with some larger measure, such as being longer. To be robust to such cases, our insight is to use the length of each medial branch to weight the SAT scaling. In order to characterize the length of a medial axis branch, we use the Extended Distance Function (EDF) proposed by Lui et al. [10].

9.5.2.1 The Extended Distance Function

In [10], the authors propose to compute at each point of the medial axis a quantity EDF (the Extended Distance Function), which measures the distance to the closest extremity of the shape corresponding to a degree-1 node on the longest path of the medial axis containing a given medial point. The set of medial balls corresponding to this longest path is called a *tube*, and EDF is the distance to the closest extremity of the *tube*. This quantity is defined for each point in the medial axis, and may be infinite if the medial axis contains loops (this happens if there are holes in $\mathcal{S}$). The authors also introduce a notion of the center of a shape based on its medial axis representation and EDF. They consider all paths along the medial axis, starting from one degree-1 node, and reaching to another degree-1 node of the medial axis. The midpoint of the path that is of maximal length is called the EMA (Extended Medial Axis). If the medial axis is acyclic (no loops), then the EMA is finite, else it is infinite. Figure 9.10 shows an example of EDF and EMA along the medial axis.

9.5.2.2 Feature Detection

Let $\mathcal{S}$ be simply connected. We propose a scaling factor based on the extended distance function. We partition the medial axis into branches. A branch is defined as a 1D subset of the medial axis on which EDF is continuous (such as straight parts in Fig. 9.10). On a branch b, we consider the factor

$$\tilde{\ell}(b) = \sup_{x \in b} \{\mathrm{EDF}(x)\}.$$

We propose to tune the scaling factor of the SAT algorithm using the branch length factor $\ell(b) \in (0, 1]$:

$$\ell(b) = \frac{\tilde{\ell}(b)}{\max_{b'} \tilde{\ell}(b')}.$$

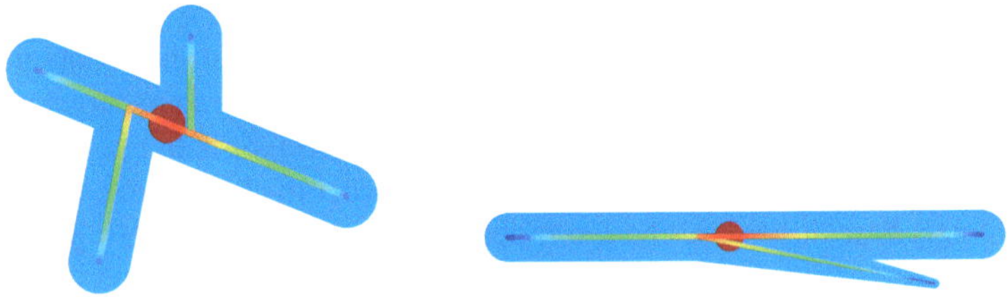

Fig. 9.10 The medial axis of two shapes, the EDF along the medial axis (in color), and the EMA (*red dot*)

This leads to the following algorithm:

Algorithm 2 The SAT-EDF feature detection algorithm

Input: Given $\mathcal{S}$, and some scaling threshold $s > 1$.
1. Compute $\mathcal{M}$, the medial axis.
2. For each branch b of the medial axis, compute $\ell(b)$, the EDF-based length.
3. For each point x_i along $\mathcal{M}$, scale the medial ball of radius $R(x_i)$ by a factor $s\,\ell(b)$, where b is the branch that contains x_i.
4. If the medial ball of x_i is contained in the union of other medial balls, label this medial point as belonging to a shape feature.

9.5.3 Area-Weighted SAT

The length is important in determining shape features in cases where the medial balls have nearly constant radii. Similarly, the area can also be important, as stated in (H4). Parts of shapes can be very long, yet have small medial radius. In these cases, the results of the user study suggest that the parts of the shape that have a smaller area are more likely to be a shape feature. For these reasons, we propose a weighted version of EDF that is a proxy for the area of a shape and can be easily computed using the medial representation. This weighted EDF is then used to adjust the scaling of the SAT algorithm to incorporate the area corresponding to a medial branch.

9.5.3.1 The Weighted Extended Distance Function

We propose an extension of EDF, weighted EDF (WEDF), for simply connected 2D shapes that takes into account not only the length of the shape but also its thickness. We can similarly define the WEMA which generalizes the EMA. WEDF is a weighted variant of EDF, and corresponds to a weighted integral of the medial radii along branches of the medial axis. For each point along the medial axis, there is a path such that the area of the union of the medial balls along this path is maximal. One can compute the area of the part of this *tube* that lies to the left and to the right of the chosen medial point. WEDF is the smaller of these two areas. Recall that EDF was similarly defined, in terms of medial length not shape area (see Sect. 9.5.2.1).

We first give some auxiliary definitions that will aid in defining WEDF. These quantities are illustrated in Fig. 9.11.

1. For any point x along the medial axis that has exactly two contact points we define two angles $\theta_1(x)$ and $\theta_2(x)$ as follows: Form a line between the two contact points and x. The angle between these lines and the tangent of the medial axis at x yield $\theta_1(x)$ and $\theta_2(x)$. Either the tangent or negative tangent can be used to calculate these angles; this leads to different values of $\theta_1(x)$ and $\theta_2(x)$,

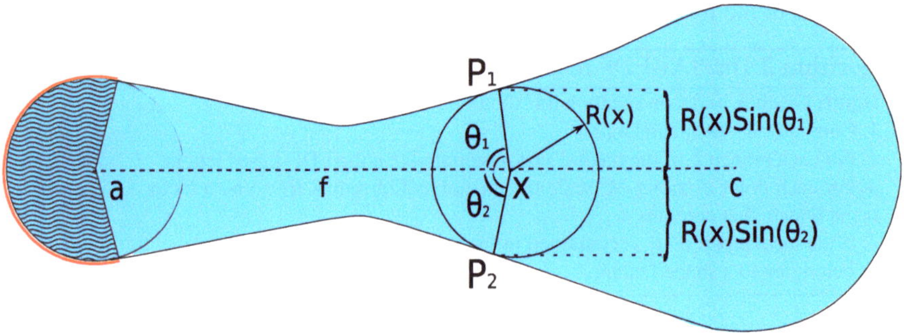

Fig. 9.11 Computation of WEDF. The *dashed area* on the left corresponds to the quantity $A_f(a)$ of the medial endpoint a. The quantities relevant to compute WEDF, $R(x)\sin(\theta_1)$ and $R(x)\sin(\theta_2)$, are illustrated

however, the same value of $\sin(\theta_1(x))$ and $\sin(\theta_2(x))$ will be obtained. The set of medial points x with more than two corresponding boundary points has measure zero, so it suffices to consider these two angles.

2. Let f be a connected curve within the medial axis, with endpoints a and c. At these endpoints, $A_f(a)$ and $A_f(c)$ refer to the area of the circle sector of the endpoint. This circle sector is defined by all boundary points of $\mathcal{S}$ that are a distance $R(x)$ from the endpoint a or c, respectively. It can be computed as follows: Compute the one-sided limits $\theta_1(a) = \lim_{x\to a}\theta_1(x)$ and $\theta_2(a) = \lim_{x\to a}\theta_2(x)$ in radians, where the tangent is calculated to point towards the extremity of the shape. Then $A_f(a) = \frac{\theta_1(a)+\theta_2(a)}{2}R(a)^2$.

See Fig. 9.11 for an illustration of θ_1, θ_2 and A_f.

Definition 9.2. Let f be any connected curve within the medial axis, and let a and c be the end points of f. We define the **weight** of f by:

$$W(f) = \int_f R(x)\,(\sin\theta_1(x) + \sin\theta_2(x))\,ds(x) + A_f(a) + A_f(c),$$

where s is the arclength parametrization of f.

This new metric integrates the radius along the path f, and leads naturally to definitions of the weighted analogs of EDF and EMA.

Definition 9.3. Let f be a simple path with endpoint a and c.

1. We first define the Weighted Extended Distance Function at a point x relative to a simple path f as:

$$\mathrm{WEDF}_f(x) = \min_{y=a,c}\left(\int_x^y R(t)\,(\sin\theta_1(t) + \sin\theta_2(t))\,ds(t) + A_f(y)\right).$$

2. We call the Weighted Extended Distance Function, or **WEDF** the following quantity:

$$\mathrm{WEDF}(x) = \sup_{f \ni x} \left\{ \mathrm{WEDF}_f(x) \right\}.$$

In this paper we do not consider shapes with holes, or equivalently, we do not consider medial axes with loops. The definition of WEDF above was made under this assumption of no loops. The measure EDF is defined in a similar way, but allows EDF to become infinite if there are loops in the medial axis [10]. We do not pursue this extension because computationally, we require that WEDF be finite—the area of the shape is finite even when there are holes. Intuitively, the WEDF measures area, whereas EDF measures the length along the medial axis.

To define the Weighted Extended Medial Axis (WEMA), consider the path $\tilde{f}$ in the medial axis of maximal weight $\tilde{f} = \operatorname{argmax} W(f)$. The **WEMA** is the point x such that $\mathrm{WEDF}_{\tilde{f}}(x) = W(\tilde{f})/2$. For an example of the computation of WEDF and WEMA and its comparison with EDF and EMA, see Fig. 9.12.

9.5.3.2 Construction of WEDF

To compute WEDF, we first discretize the medial axis and represent it as a weighted graph. Then, starting from degree-1 nodes, we compute WEDF and work inwards along the medial axis, updating WEDF for each new discretized medial point. At each junction, we choose the maximum previous WEDF value to proceed. This algorithm is analogous to the scheme for computing EDF [10]. The discrete algorithm is summarized below:

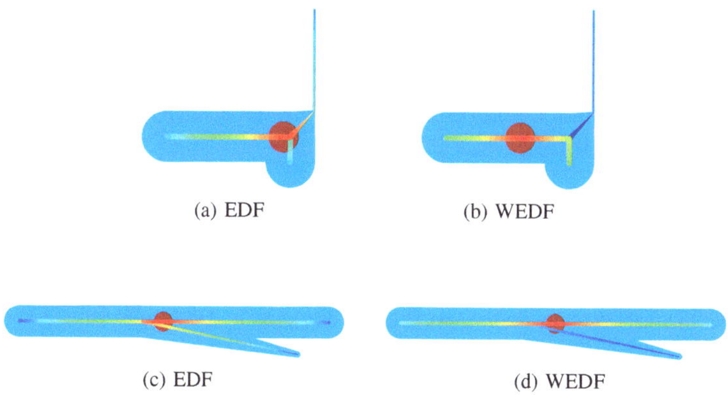

Fig. 9.12 Comparison of EDF (**a**) and WEDF (**b**) for a shape with a thin feature and for a shape with a slowly-changing radius (**c**)–(**d**)

1. Initialize each degree-1 node x_i with $\mathrm{WEDF}(x_i) = A_f(x_i)$ where f is a path bearing the node x_i. Other nodes are initialized with infinite WEDF.
2. For each degree-1 node x_i, if its neighbor x_j is not degree-1, update

$$\mathrm{WEDF}(x_j) = \min\left(\mathrm{WEDF}(x_j), \mathrm{WEDF}(x_i) + |x_i - x_j| * \frac{g(x_i) + g(x_j)}{2}\right),$$

   where $g(x) = R(x)\,(\sin\theta_1(x) + \sin\theta_2(x))$.
3. At any junction, wait until there is only one adjacent medial point with infinite WEDF. When this is true, update the junction point using the largest of the finite adjacent WEDF values (and the formula in step 2).
4. Terminate the algorithm when all discrete medial points have been visited (equivalently, when all WEDF values are finite).

9.5.3.3 Feature Detection

Assume that $\mathcal{S}$ is simply connected. We propose a scaling factor based on WEDF that is analogous to the EDF-based scaling. We partition the medial axis into branches. A branch is defined as a 1D subset of the medial axis on which WEDF is continuous. On a branch b, we consider the factor

$$\tilde{\omega}(b) = \sup_{x \in b}\{\mathrm{WEDF}(x)\}.$$

We propose to tune the scaling factor of the SAT algorithm using the branch area proxy $\omega(b) \in (0, 1]$:

$$\omega(b) = \frac{\tilde{\omega}(b)}{\max_{b'} \tilde{\omega}(b')}.$$

This leads to the generalization of the Scale Axis Transform for feature detection described in Algorithm 3. For a comparison of the SAT, SAT-EDF and SAT-WEDF algorithms, see Fig. 9.13.

Algorithm 3 The SAT-WEDF feature detection algorithm

Input: Given $\mathcal{S}$, and some scaling threshold $s > 1$.
1. Compute $\mathcal{M}$, the medial axis.
2. For each branch b of the medial axis, compute $\omega(b)$, the WEDF-based area.
3. For each point x_i along $\mathcal{M}$, scale the medial ball of radius $R(x_i)$ by a factor $s\,\omega(b)$, where b is the branch that contains x_i.
4. If the medial ball of x_i is contained in the union of other medial balls, label that a medial point corresponding to a shape feature.

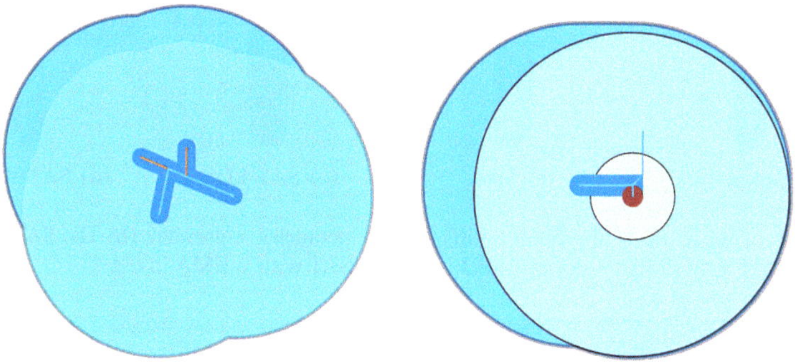

Fig. 9.13 These figures show dilations for the SAT (*darker blue* in the back) and SAT-EDF algorithms (*light blue*). On the *left figure*, since the radius is similar on all branches, the SAT will not detect any features. For SAT-EDF, the dilation is decreasing, and therefore the *yellow* parts of the medial axis are eventually considered features of the main shape (the *blue* part of the medial axis). On the *right*, the SAT dilation is indicated in the back as well, and the other regions correspond only to the dilation of the *red point*, for SAT-EDF (the color is the same *blue* as on the *left figure*) and SAT-WEDF (give by a lighter color), respectively. For SAT-EDF, we can see that the shape is less dilated on the parts that are not labeled as the main axis by EDF, and is rapidly labeled as a feature of the main shape by the dilation process. This is not the case for the SAT-WEDF region corresponding to the red point—it will never be eaten up since it belongs to the biggest branch in terms of area. This figure explains the results of Fig. 9.19

9.6 Algorithm Results and Comparisons

In this section, we compare the four algorithms we have discussed for detecting shape features—the geometric algorithm (cf. Algorithm 1), the standard Scale Axis Transform, the SAT-EDF algorithm (cf. Algorithm 2), and the SAT-WEDF algorithm (cf. Algorithm 3). Parts of the medial axis corresponding to the identified features are highlighted in yellow, whereas the part of the axis corresponding to the main shape are in blue. The geometric algorithm was based on local medial information—radius and orientation of the medial branches. The SAT algorithm was the medial pruning method that was the inspiration for the algorithms proposed [6]. As the Scale Axis Transform is not able to detect shape features when the medial radii are nearly constant, SAT-EDF corrects for this by using branch length in the SAT scaling. Similarly, to incorporate both length and thickness, the SAT-WEDF algorithm uses a proxy for branch area in the SAT scaling. The shapes on which we test the algorithms are similar to the shapes in the user study and have only the main shape and one level of features to simplify the SAT-based results. Below, we compare the results of these algorithms on a series of test shapes, and highlight the strengths and weaknesses of each approach.

Note that for some of our tests (e.g., in Fig. 9.10), we used a simplified skeletal representation instead of the exact medial axis. We did not notice any change in the quality of results.

(a) Geometric (b) SAT (c) SAT-EDF (d) SAT-WEDF

Fig. 9.14 Comparison of algorithm results. (**a**) The geometric algorithm, (**b**) The SAT, (**c**) EDF-adjusted SAT with EMA in *red*, (**d**) WEDF-adjusted SAT with WEMA in *red*

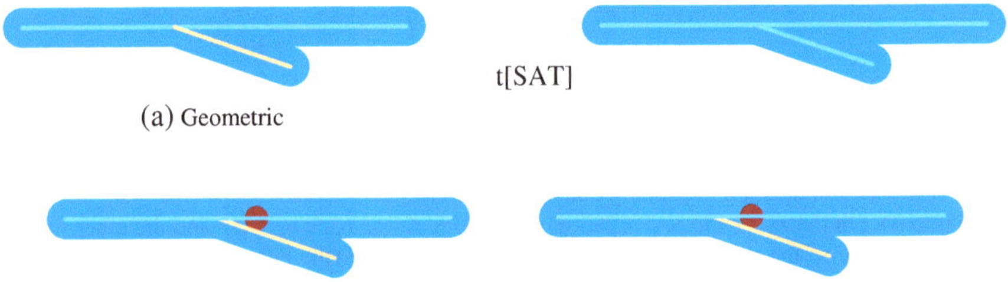

(a) Geometric t[SAT]

(c) SAT-EDF (d) SAT-WEDF

Fig. 9.15 Comparison of algorithm results. (**a**) The geometric algorithm, (**b**) The SAT, (c) EDF-adjusted SAT with EMA in *red*, (**d**) WEDF-adjusted SAT with WEMA in *red*

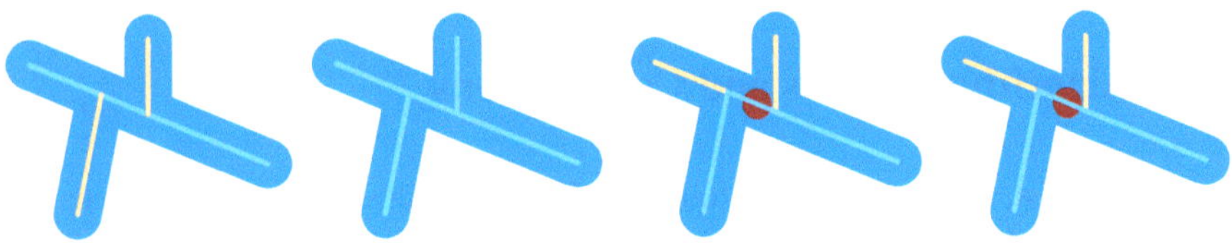

(a) Geometric (b) SAT (c) SAT-EDF (d) SAT-WEDF

Fig. 9.16 Comparison of algorithm results. (**a**) The geometric algorithm, (**b**) The SAT, (c) EDF-adjusted SAT with EMA in *red*, (**d**) WEDF-adjusted SAT with WEMA in *red*

In Fig. 9.14, the results of all four algorithms are shown on a shape with a thin shape feature. In this case, all algorithms are consistent, and correctly identify the thin branch as the shape feature.

In Fig. 9.15, the results of the algorithms are shown on a different shape. In this case, the geometric algorithm, SAT-EDF and SAT-WEDF agree, and correctly identify the shorter branch as a shape feature. However, the SAT algorithm cannot identify the smaller branch because the shape has near-constant thickness. This shows a case where the three algorithms proposed here outperform the SAT.

Figure 9.16 depicts a case where the geometric algorithm outperforms the SAT-based algorithms. The geometric algorithm chooses the shape features which are most consistent with the user study (see Fig. 9.3). The SAT algorithm does not detect shape features due to the nearly constant medial radius of the shape. The SAT-EDF

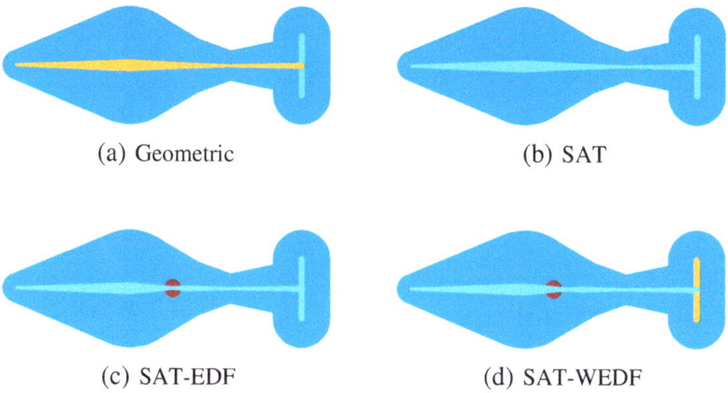

Fig. 9.17 Comparison of algorithm results. (**a**) The geometric algorithm, (**b**) The SAT, (**c**) EDF-adjusted SAT with EMA in *red*, (**d**) WEDF-adjusted SAT with WEMA in *red*

and SAT-WEDF algorithms produce the same results (due to the near-constant medial radius), choosing the longest path as the main shape and the smaller branches as features, neglecting orientation.

On Fig. 9.17 however, the local character of the geometric algorithm leads to a non-intuitive choice for the feature: the aligned segments are chosen as the main shape, despite their small size compared to a large feature. The SAT-based algorithms are able to identify the larger left part of the shape as part of the main shape, because they employ a global strategy to identify shape features. However, here all SAT-based algorithms use the same scaling factor: whereas SAT-WEDF is already able to identify the right part as the feature, the SAT and SAT-EDF algorithms do not. Eventually, for larger scaling, both the SAT and SAT-EDF algorithms identify the same part as a feature.

In Fig. 9.18, the results are shown for two shapes with repeated features. The SAT algorithm is unable to detect shape features due to the nearly-constant radius of the medial balls for these two shapes. However, the geometric algorithm as well as SAT-EDF and SAT-WEDF both detect the same shape features. For the shape with less protruding shape features, these results are consistent with the user study. For the shape in the first row, the end features are not detected. None of these algorithms currently detect repeated shape features. In the first row of Fig. 9.18, the two end segments may not be considered shape features without the middle three features. This leads us to a possible extension of the current methods, based on the following statement:

Repeated parts of a shape are likely to be perceived as shape features.

In Fig. 9.19, an example is shown where SAT-WEDF outperforms SAT-EDF. In this case, the geometric algorithm, the SAT algorithm, and SAT-WEDF all correctly identify the long, thin segment as the shape feature. For SAT-EDF, the algorithm

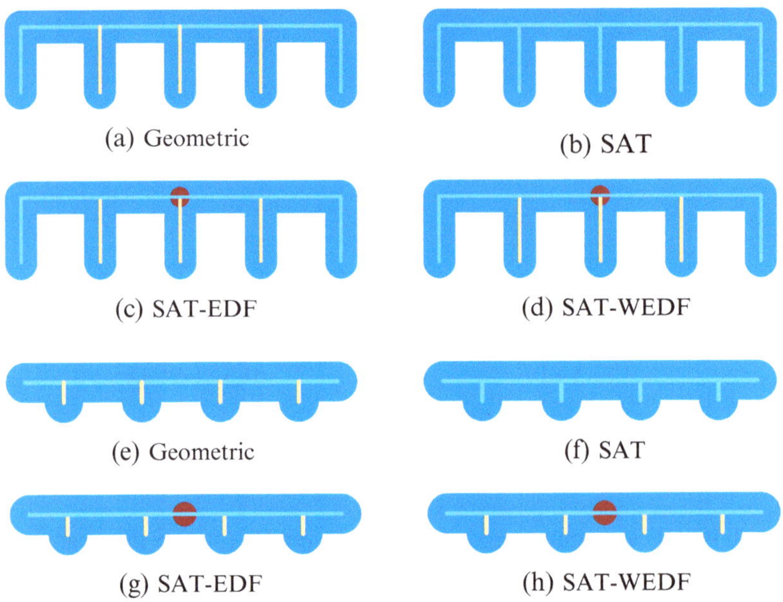

Fig. 9.18 Comparison of algorithm results on two shapes. (**a**) and (**e**) The geometric, (**b**) and (**f**) the SAT, (**c**) and (**g**) EDF-adjusted SAT with EMA in *red*, (**d**) and (**h**) WEDF-adjusted SAT with WEMA in *red*

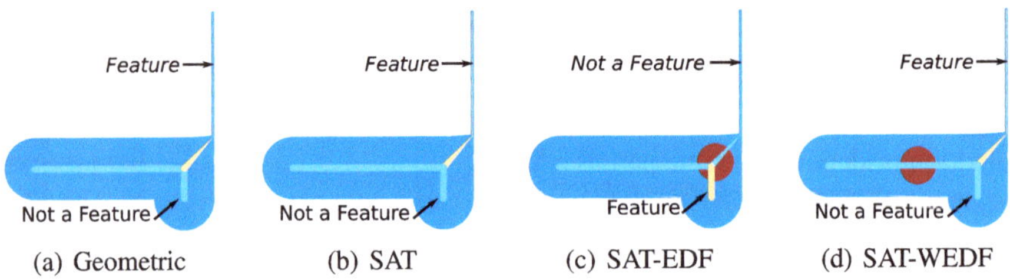

Fig. 9.19 Comparison of algorithm results. (**a**) The geometric algorithm, (**b**) The SAT, (**c**) EDF-adjusted SAT with EMA in *red*, (**d**) WEDF-adjusted SAT with WEMA in *red*

considers this long thin feature to be the main shape, whereas the nob at the bottom was identified as a feature. As this was considered to be part of the main shape in the user study, these results show the sensitivity of SAT-EDF to long, thin shape parts.

These results demonstrate the strengths and weaknesses of four algorithms we tested for detecting shape features. Our conclusions are given next.

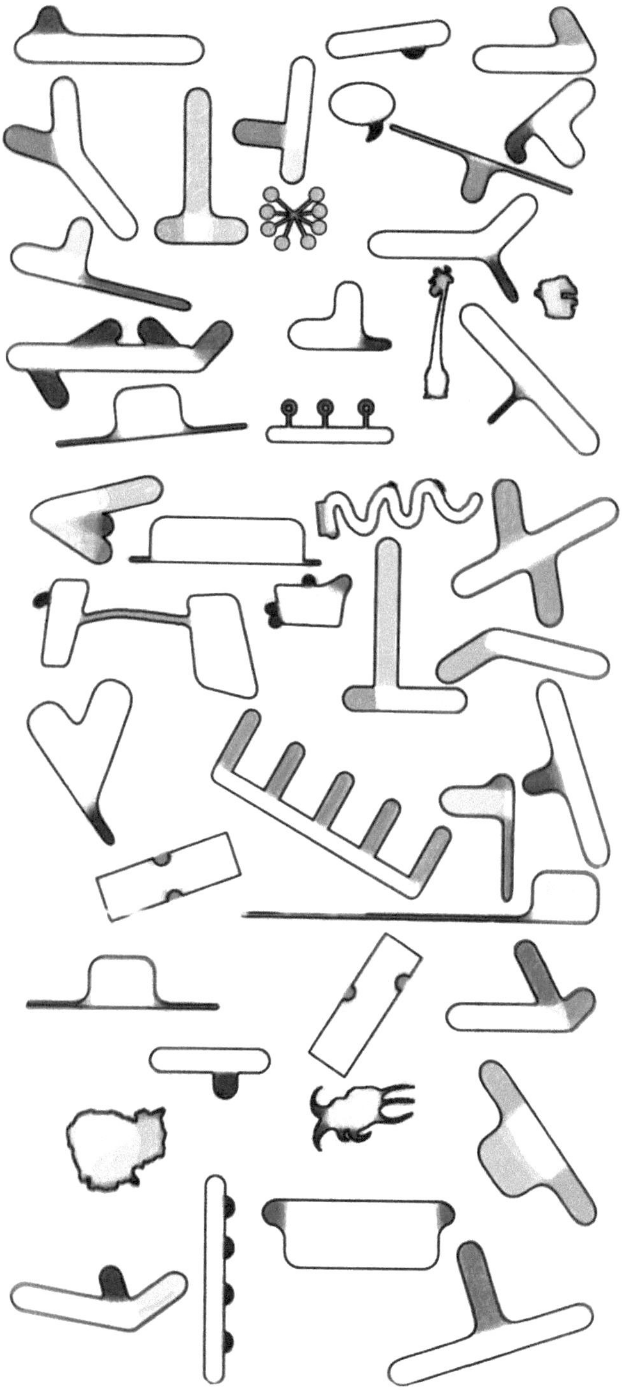

Fig. 9.20 Results of the user study. This shows the average pixel value for all users on all 44 shapes administered

9.7 Discussion and Conclusion

In this paper, the perception of 2D shape features was measured via a user study. This information was then used to construct three algorithms to identify shape features. The first algorithm was a geometric algorithm based on change in radius and orientation at junctions of the medial axis. This method offered results that were consistent with the user study on most examples, but may fail as it is only based on local criteria (see Fig. 9.17). The Scale Axis Transform uses global criteria and performs well on many examples, but cannot identify features correctly on shapes with near-constant thickness, whereas the user study shows very clear identification of features on such shapes. We thus extended the SAT algorithm to incorporate branch length (based on EDF) in the scaling. This method—SAT-EDF—is able to identify features on a shape with near-constant thickness. However, the lengthier branches are always considered the main shape, which is not consistent with the user study. The last algorithm—SAT-WEDF—was able to handle cases with long and thin features, as it uses both length and thickness information in the SAT scaling. In this case, the last algorithm reproduced well the results of the user study. However, the results of these algorithms on a set of test shapes indicated that all algorithms should be coupled with a mechanism for detecting repeated shape features.

The benefit of using the SAT-based algorithms is that they immediately offer a hierarchy of shape features by tuning the scaling threshold s. The shapes in this paper were chosen to have only two level of details (main shape and features) in order to identify perceptual features as defined by a tractable user study. These two levels are adapted for handling shape deformation operations: properties of the features may be preserved, while the main shape undergoes the deformation. However, in future work we also intend to benefit from the multiresolution inherent to the SAT based algorithms. By choosing an appropriate scaling factor, s, these algorithms automatically find the features detected under that scaling. Determining the scale of details on shapes is well-suited to be investigated in an expanded perceptual study.

References

1. Amenta, N., Choi, S., Kolluri, R.K.: The power crust. In: Proceedings of the Sixth ACM Symposium on Solid Modeling and Applications, Ann Arbor, pp. 249–266. ACM (2001)
2. Attali, D., Montanvert, A.: Computing and simplifying 2D and 3D continuous skeletons. Comput. Vis. Image Underst. **67**(3), 261–273 (1997)
3. Blum, H., Nagel, R.N.: Shape description using weighted symmetric axis features. Pattern Recognit. **10**(3), 167–180 (1978)
4. Blum, H., et al.: A transformation for extracting new descriptors of shape. Models Percept. Speech Vis. Form **19**(5), 362–380 (1967)
5. Chazal, F., Lieutier, A.: The λ-medial axis. Graph. Models **67**(4), 304–331 (2005)
6. Giesen, J., Miklos, B., Pauly, M., Wormser, C.: The scale axis transform. In: Proceedings of the 25th Annual Symposium on Computational Geometry, Aarhus, pp. 106–115. ACM (2009)

7. Kimia, B., Tannenbaum, A., Zucker, S.: Toward a computational theory of shape: an overview. In: Faugeras, O. (ed.) Computer Vision – ECCV 90, Antibes. Volume 427 of Lecture Notes in Computer Science, pp. 402–407. Springer, Berlin/Heidelberg (1990)
8. Kimia, B.B., Tannenbaum, A.R., Zucker, S.W.: Shapes, shocks, and deformations I: the components of two-dimensional shape and the reaction-diffusion space. Int. J. Comput. Vis. **15**(3), 189–224 (1995)
9. Leyton, M.: Symmetry-curvature duality. Comput. Vis. Graph. Image Process. **38**(3), 327–341 (1987)
10. Liu, L., Chambers, E.W., Letscher, D., Ju, T.: Extended grassfire transform on medial axes of 2D shapes. Comput. Aided Des. **43**(11), 1496–1505 (2011)
11. Miklos, B., Giesen, J., Pauly, M.: Discrete scale axis representations for 3D geometry. ACM Trans. Graph. **29**(4), 101 (2010)
12. Prasad, L.: Rectification of the chordal axis transform and a new criterion for shape decomposition. In: Discrete Geometry for Computer Imagery, Poitiers, pp. 263–275. Springer (2005)
13. Sebastian, T.B., Klein, P.N., Kimia, B.B.: Recognition of shapes by editing shock graphs. In: International Conference on Computer Vision, Vancouver, vol. 1, pp. 755–762 (2001)
14. Shaked, D., Bruckstein, A.M.: Pruning medial axes. Comput. Vis. Image Underst. **69**(2), 156–169 (1998)
15. Sud, A., Foskey, M., Manocha, D.: Homotopy-preserving medial axis simplification. Int. J. Comput. Geom. Appl. **17**(05), 423–451 (2007)
16. Tagliasacchi, A.: Skeletal representations and applications (2013). arXiv preprint arXiv:1301.6809
17. Tam, R., Heidrich, W.: Feature-preserving medial axis noise removal. In: Computer Vision – ECCV 2002, Copenhagen, pp. 672–686. Springer (2002)
18. Zeng, J., Lakaemper, R., Yang, X., Li, X.: 2D shape decomposition based on combined skeleton-boundary features. In: Advances in Visual Computing, Las Vegas, pp. 682–691. Springer (2008)

MIX
Papier aus verantwortungsvollen Quellen
Paper from responsible sources
FSC® C105338

If you have any concerns about our products,
you can contact us on
ProductSafety@springernature.com

In case Publisher is established outside the EU,
the EU authorized representative is:
Springer Nature Customer Service Center GmbH
Europaplatz 3, 69115 Heidelberg, Germany

Printed by Libri Plureos GmbH
in Hamburg, Germany